THE INFANCY OV THOUGHT

JACOB A. EDER

THE INFANCY OV THOUGHT

COPYRIGHT PAGE

© 2024 BY JACOB A. EDER

ALL RIGHTS RESERVED. NO PART OF THIS BOOK, INCLUDING THE ORIGINAL POEMS, STORIES, AND COPYRIGHT-PROTECTED WORKS OF ORIGINAL ART MAY BE REPRODUCED, DISTRIBUTED, OR TRANSMITTED IN ANY FORM OR BY ANY MEANS, INCLUDING PHOTOCOPYING, RECORDING, OR OTHER ELECTRONIC OR MECHANICAL METHODS, WITHOUT THE PRIOR WRITTEN PERMISSION OF THE PUBLISHER, EXCEPT FOR BRIEF QUOTATIONS EMBODIED IN CRITICAL REVIEWS AND CERTAIN NONCOMMERCIAL USES PERMITTED BY COPYRIGHT LAW. FOR PERMISSION REQUESTS, PLEASE REACH OUT TO THE AUTHOR @TERMINUSGENESIS@GMAIL.COM, OR FIND HIM ON SOCIAL MEDIA PLATFORMS, UNDER THE BUSINESS ENTITY T3TINNOVATIONS, OR BY NAME.

THE INFANCY OV THOUGHT

ISBN: 9798343645682

PRINTED IN THE UNITED STATES OF AMERICA

FIRST EDITION

CONTENTS

CONTENTS 3

CH. 1: B.D. LIFE BEFORE DARWIN 1

CH. 2: D.D. DURING DARWIN 99

CH. 3: A.D. AFTER DARWIN 126

CH. 4: SOCIAL UTILITY 187

CH. 5: NATURE 227

CH. 6: NURTURE 240

CH. 7: BIOLOGICAL PRINCIPLES 261

CH. 8: UNIFIED THEORY OF BIOLOGICAL AND NEUROPSYCHOLOGICAL SCHOOLS OF THOUGHT [EDERIAN EVOLUTION] 273

THE INFANCY OV THOUGHT

DEDICATION

For the fallen, for my beloveds…

For all those who did not live to see my long-anticipated literary debut published. Your memories dwell in every word, your spirits imbue every thought. This is for you, now and always.

THE INFANCY OV THOUGHT

FOREWORD

The cosmos is not a mere collection of stars and planets, nor a passive repository of galaxies and matter. It is an endless symphony of forces—unseen, unfathomable, and terrifyingly beautiful—whirling through the void, playing out the melody of existence in a dance as old as time itself. But even before time, there was something older: the abyss, vast and cold, where silence reigned in its own cruel majesty. It is from this abyss, from this fathomless and incomprehensible void, which all things were born—the violent genesis of creation itself.

In these pages, we confront the tumultuous origins of the universe, an epic unfolding from nothingness to something, from chaos to order, and from life to death. The birth of existence was not a delicate unveiling; it was a cataclysmic eruption. The first pulse of being was violent, the first tremor of creation shook the fabric of what could be imagined, and from this primordial chaos, the stars flared into existence—bright and fierce, harbingers of both life and destruction.

But creation is not an act of singular purpose. It is an ongoing, cyclic affair—one marked by endless birth, death, and rebirth. It is the story of dust and fire, of cold and warmth, of light and dark. The endless turmoil of becoming is embedded in every atom, every quark, every element which shapes our reality. For within the stars lies the power to create life, yet within them also resides the seed of entropy, the inevitability of collapse. This paradox—which from death comes life, and from life, death—remains the ultimate truth that binds the universe together.

The Infancy Ov Thought is not merely a narrative of scientific inquiry; it is an exploration of the very heart of existence. It is a meditation on the cyclical nature of all things—where the boundaries between science and philosophy dissolve, and where physics itself becomes an expression of the profound mysteries of life. This work is not just a chronicle of cosmic phenomena but a reflection on the forces that both shape and unravel our universe. It is a contemplative journey, yearning to understand humanity's fleeting existence, forged in the furnace of the stars, and ultimately swept away by the inexorable march of time.

In this collection of thoughts, concepts, and hypotheses, we delve into the chaos from which order arises, and the inevitable decay that follows. Here, the universe is not a distant, cold machine but a living entity—an ever-unfolding process of creation, adaptation, and destruction. We are not separate from the forces that govern the cosmos; we are part of them. Our very being is a result of the same forces that guide the stars, the same laws that govern the galaxies, and the same energy that propels all matter from the smallest atom to the largest nebulae.

Through these pages, we journey into the depths of the unknown. The questions posed within are not those of certainty, but of awe. What is the nature of life, time, and consciousness? What compels the universe to form galaxies, to breathe life into what was once inert? And what drives the cosmos toward its eventual entropy and decay? These are not merely scientific questions—they are the very essence of human curiosity.

The act of creation, whether it be the birth of stars or the emergence of life, is not an event isolated from us. It is an intimate process, one that we are bound to as participants and witnesses. We are but a fleeting moment in the universe's grand cycles of birth, existence, and destruction, and this work seeks to examine those moments with reverence and reflection.

This is not simply a scientific treatise. It is a reckoning with the profound truths of existence. It reflects on the origins of all things and the inescapable reality of their end. It is about us—the fragile custodians of consciousness—gazing into the void, searching for meaning in a universe that often seems indifferent to our existence. Yet, through this search, we come to understand that we are not merely passive observers; we are part of the universe's unrelenting drive to create and to destroy.

The Infancy Ov Thought invites readers to witness the universe in all its beauty and terror. In its grand cycles, we see a reflection of our own lives—emerging from the depths of the unknown, striving for meaning, and ultimately returning to the void. This is not a story of certainty, but of contemplation. It is about the delicate balance between creation and destruction, a balance that we, as sentinels of consciousness, are both a product of and participants in.

This, then, is The Infancy Ov Thought —a reflection on the profound and relentless forces that shape not only the cosmos but our very existence.

THE INFANCY OV THOUGHT

PREFACE

At the tender age of fifteen, when the mind is still in the throes of developmental transitions and the world is an uncharted expanse of untapped knowledge, my earliest reflections upon life, science, and the intricacies of the human condition took form. The Infancy Ov Thought represents not only the raw musings of a young mind grappling with complex philosophical and scientific questions but also serves as a testament to the enduring power of youthful curiosity.

This work, nearly untouched since its original inception, is a window into the formative intellectual period of my life, offering insights into concepts that would later crystallise into fully developed hypotheses and scientific inquiry. What stands before you is a collection of thoughts, hypotheses, and writings from a fifteen-year-old who dared to question, theorise, and imagine beyond the traditional boundaries of knowledge.

The essence of this book lies not only in the novelty of thought but also in the willingness to embrace complexity without fear. To delve into areas of intellectual exploration that even seasoned academics often shy away from, I challenged myself to explore and integrate scientific paradigms with the raw emotional landscape that defines adolescence.

THE INFANCY OV THOUGHT

Readers will find a blend of nascent scientific hypotheses, embryonic theoretical frameworks, and evolving philosophical inquiry, all stemming from an adolescent perspective. And while these ideas were conceived in the crucible of youthful naiveté, the depth of investigation and the breadth of questions posed provide a unique lens through which to view the development of intellectual thought.

It is important to understand that this book serves as a foundation—a cornerstone of intellectual curiosity—written at an age when one is acutely aware of both the limitations of their knowledge and the boundless potential to expand it. I offer this work not merely as a recollection of youthful academic pursuit but as a reminder that the journey of discovery is continuous and that even in our early stages of thought, profound ideas can emerge.

This work is not only a testament to the intellectual struggles I faced but also a reflection of the complexities I sought to unravel. It was an attempt to give order to the chaos, to understand the unknown, and to make sense of the world from an evolving, self-taught, and autodidactic perspective.

I present this to you as a glimpse into the mind of a teenager who was unwilling to accept the limits of conventional understanding—who sought to go beyond and pave the way for deeper, more intricate explorations of science, philosophy, and existence itself.

THE INFANCY OV THOUGHT

... AND SO, IT BEGINS...

CH. 1: B.D. LIFE BEFORE DARWIN

THE VOID AND PRECREATION

'WILLED MOTION,' (THE BANG & THE SPARK)

ABIOGENESIS (FROM ATOM, CAME ADAM) & PRESCIENCE

THE INFANCY OV THOUGHT

THE VOID & PRECREATION

It is enthralling to ponder the vastness of existence, to feel the weight of how precious life truly is. It mesmerises us, as we attempt to grasp even a fraction of its infinite mystery. Our brains, much like grand libraries or interconnected transit systems, store knowledge and experiences, each neuron a node in a complex network. And yet, in the deepest recesses of our minds, something far more profound is at play—something that ties the abstract beauty of consciousness, *Willed Motion*, and emotional intelligence together into the vast tapestry of existence. We are beings designed, whether by evolution or divinity, to feel deeply, to love, and to long for love. It is autonomic, instinctual—an inescapable truth that speaks to something beyond the mere mechanical functions of our bodies.

To merely live, to feel the cold wind on our skin, to taste the bitter tang of sorrow or the sweet rapture of joy—these experiences are miraculous in their own right. The very fact that we exist, that we can perceive, think, and feel, is an anomaly in the grand, seemingly indifferent expanse of the universe. In this, we find a call to action, a reminder not to squander this fleeting opportunity to live. Life, in its unfathomable complexity, is far too improbable, too mathematically impossible, to be taken for granted.

Consider the magnitude of chance, the infinitesimal odds that out of the infinite possible configurations of matter, we were the ones to emerge. To witness a sunrise, to hear a symphony, to hold a loved one close—these experiences defy the cold calculus of random chance. They beckon us to consider the possibility that something greater exists, that our existence is not mere accident but part of a grand, deliberate motion—what I have termed *Willed Motion*.

From the moment of the universe's birth, when that initial act of volition sparked the formation of time, space, and matter, there has been a driving force, a cosmic will, behind every movement, every evolution, every beat of the human heart. It is a force that binds us to the universe and to each other, for we are not just physical entities; we are conscious beings, capable of introspection, love, and empathy. This conscious capacity is not merely a byproduct of biological evolution, but rather a reflection of that initial act of volition—the *Willed Motion* that gave rise to all things.

In contemplating emotional intelligence, we find that it is not an isolated phenomenon but deeply interconnected with our broader understanding of the universe. Emotional intelligence—our ability to navigate abstract concepts, to discern right from wrong, to filter beauty from filth—is an echo of this cosmic consciousness. It is as if we are wired to perceive and interpret the world not merely through our senses but through the lens of our souls, which resonate with the *Willed Motion* that sustains the cosmos. This intelligence allows us to navigate the complexities of life, to wrestle with moral dilemmas, and to seek meaning beyond mere survival.

We may not have found what is traditionally called the "soul," but we do know that consciousness resides

THE INFANCY OV THOUGHT

within us—specifically in the frontal lobe, where logic, individuality, and a sense of self reside. It is here, within the neural web of our brains, that we experience the full spectrum of human emotions, from the deepest sorrow to the highest joy. When we feel threatened, doomed, or bereft, our bodies respond in kind, imitating physiological pain. This suggests that our emotions, complex and often confounding, are not random but tied intrinsically to the fabric of who we are—beings capable of perceiving more than just the physical world.

Yet this capacity for emotional depth, for empathy and understanding, begs a larger question: why? Why do we long, love, and hate? Why do we feel so intensely? It seems absurd to reduce these profound experiences to mere biological imperatives. Instead, it is far more reasonable to consider that there is something greater at work—a force, a reason, a purpose for our existence. This force, which I believe to be the same *Willed Motion* that sparked the universe into being, is what drives us to seek connection, to yearn for something beyond ourselves.

To suggest that life is merely a series of random events, devoid of meaning, is to miss the larger picture. Life's intricacies, its beauty, and its tragedy are not mere accidents. The very fact that we can feel deeply, and that we can love, despair, and hope, points to a greater design. To think that all of this—the joy, the pain, the endless search for meaning—ends in nothingness is a tragic notion. Surely, there must be more.

When we contemplate the idea of death, the notion that all our experiences, all our memories and emotions, could fade into nothingness is deeply unsettling. It feels wrong, as if we are meant for more than this fleeting existence. And perhaps we are. Perhaps, as I suggest in the framework of *Willed Motion*, there is something—someone—waiting for us on the other side of this existence. Perhaps, after our bodies cease to function, our consciousness—the part of us tied to that initial spark of creation—will be received by the same force that set the universe into motion.

In this, I plead with you to consider: what if there is more? What if our lives, with all their trials and tribulations, are not meaningless but part of a greater cosmic design? To believe in something more, in a force greater than ourselves, is not just a comforting thought but a necessary one. For without hope, without the belief in something beyond our material existence, despair inevitably follows. A world devoid of belief, devoid of hope, is one that is vulnerable to nihilism, to the cold, indifferent void that so many fear.

It is no wonder that so many cling to the idea of a higher power, of a grand design, for without it, the world becomes a far bleaker place. To deny hope is to deny what makes us human—our capacity for belief, for love, for wonder. And though scepticism and doubt are important for intellectual growth, to completely abandon the notion of something greater is to open the door to despondency and despair.

In the end, it is belief—belief in something more, something greater—that sustains us. Whether we call it *God*, *Quantum Consciousness*, or simply *Willed Motion*, the fact remains that there is something driving us, something that gave rise to the universe and all within it. And in that force, in that spark of volition, we find not just the origin of the cosmos but the origin of

THE INFANCY OV THOUGHT

ourselves, of our consciousness, and of our capacity for love, for pain, and for hope.

What I am attempting to articulate ventures into the deepest recesses of abstraction. It is a concept so vast, so overwhelmingly intricate, that it seems almost unattainable in thought—but it is within this abyss of intellectual pursuit that the core of *Willed Motion* lies. Imagine, if you will, an expanse of cold, dark, and infinite vastness—a place where time does not pass, where space does not stretch, and where existence itself remains in a boundless, dimensionless state. In this primordial void, a single, absolute force dwells, ageless and eternal, without beginning and without end. This entity—God, the creator, or the *Quantum Consciousness* as I hypothesise—exists alone in this infinite, desolate expanse. Can you fathom the weight of such a being's sorrow, of such loneliness, borne out of timeless aeons of solitary existence?

Now, in this place of endlessness and darkness, consider the necessity of creation. Would it not be inevitable that such a being—immersed in profound solitude—would yearn for more? In this contemplation, we find the origin of *Willed Motion*, the very first act of creation, born from the necessity to exist beyond the void. It is within this sorrow, this contemplation, that the first flicker of volition ignites, propelling the universe into motion. What was once nothing became something—the first act of divine will. This is not mere chance, nor a product of randomness, as classical cosmology might suggest. It is an intentional eruption, an act of volition that gave birth to time, space, matter, and consciousness.

THE INFANCY OV THOUGHT

In this framework, the universe did not spring from a void devoid of intention, but from a prescient, conscious substrate awaiting a catalyst. The Big Bang—often viewed as the genesis of all things—is but the first expression of *Willed Motion*, the primal command that transformed the darkness into the light, the nothingness into being. This is where the hypothesis diverges from classical cosmological models, such as the Hartle-Hawking no-boundary proposal, which frames the universe's emergence as a probabilistic event. I argue that the emergence of the cosmos was not driven by mere quantum fluctuation but by conscious volition. *Willed Motion* did not arise from chance; it was a deliberate act—a metaphysical command that transcended the physical laws of quantum mechanics, sparking the formation of space-time.

This *Willed Motion*—this singular act of divine volition—is the foundation for all that exists. It is the progenitor not only of matter and energy but of consciousness itself. The universe's self-aware quantum field, which I term *Quantum Consciousness*, emerged from the collapse of quantum uncertainty into an intentional act of creation. This volition, once set into motion, continues to drive the expansion of the universe, weaving through dimensions beyond our comprehension, permeating the very structure of existence.

Herein lies the connection between *Willed Motion* and the multidimensional nature of the cosmos. Drawing from higher-dimensional theories like String Theory and *M-Theory*, which posit up to eleven or more dimensions, my hypothesis suggests that *Willed Motion* does not confine itself to our three-dimensional reality but permeates the higher-dimensional planes of existence. In these higher realms, imperceptible to the limited human senses, the initial union of positive and negative charges during the cosmic birth event continues to resonate, shaping time and matter not only in our dimension but across the many layers of reality. These higher-dimensional forces—echoes of the first *Willed Motion*—are the engines of cosmic evolution, guiding the universe through cycles of creation, transformation, and destruction.

Just as consciousness permeates the human mind, so too does it permeate the universe itself. This consciousness—this *Quantum Consciousness*—is fundamental to the fabric of reality. It is not an emergent property of biological systems but an inherent force within the cosmos. My framework aligns with the Orch-OR theory of Penrose and Hameroff, which postulates that quantum processes within the brain's microtubules give rise to consciousness. However, I extend this hypothesis further, arguing that consciousness is not confined to the brain but is embedded in the very essence of the universe, woven into the quantum structure of space-time.

In this light, the soul—the elusive entity that so many have searched for—may well be synonymous with *Quantum Consciousness*. The human mind, with its capacity for emotional intelligence, abstract reasoning, and moral judgement, is but a microcosmic reflection of the greater cosmic consciousness. Emotional intelligence itself—our ability to empathise, to discern good from evil, to navigate the complexities of ethical dilemmas—is not merely an evolutionary adaptation for survival. It is an expression of *Willed Motion* manifested within us, a reflection of the same volition that drives the universe.

This emotional intelligence, intertwined with our ability to reason, create, and philosophise, mirrors the cosmic processes that shape stars, galaxies, and black holes. The experiences of sorrow, joy, pain, and longing that we ascribe to the soul are expressions of a far greater force. When we feel love, betrayal, or grief, we are not merely responding to biological imperatives; we are resonating with the cosmic forces that birthed the universe. Our emotional depth—our capacity for abstract thought and moral reasoning—ties us to the greater *Quantum Consciousness*, to the *Willed Motion* that first ignited creation.

This is why I posit that even the speed of light, the supposed ultimate limit of physical reality, is pliable under the influence of consciousness. The peculiar phenomenon of radio-frequency travel, where a voice escapes the decay of entropy, illustrates that consciousness—intertwined with *Willed Motion*—can influence the very fabric of time and space. Consciousness can, in rare and exceptional cases, bend the physical laws of the universe, transcending the limitations imposed by light speed. In this way, our conscious minds, like the universe itself, are capable of interacting with the fundamental forces that govern reality.

Moreover, black holes, traditionally viewed as destructive forces, are recast in this hypothesis as engines of cosmic evolution. These singularities are not endpoints but reset points—junctions where *Willed Motion* exerts its most profound influence, breaking down space-time and reconfiguring matter into new forms. Within these singularities, the cyclical nature of the universe becomes apparent. Matter is compressed, broken down, and reborn, continuing the cosmic dance of creation and destruction. Black holes, in this sense, are expressions of the same *Willed Motion* that initiated the universe's birth, ensuring that the cycle of cosmic evolution remains unbroken.

The abstraction of music as the universal language, as I have articulated previously, is another reflection of this grand cosmic principle. Music transcends words, transcends culture, and speaks directly to the emotive and intellectual capacities of the human mind. It is an expression of *Willed Motion*, resonating through time and space, connecting individuals to the greater universal consciousness. Music, like consciousness, is a manifestation of the cosmic forces that shape reality.

Willed Motion is the foundational force behind both the physical and metaphysical aspects of existence. It initiated the universe's birth, drives its expansion, and shapes the evolution of consciousness. Emotional intelligence, evolutionary adaptation, and human creativity are all expressions of this primal force, guiding us toward greater understanding and transcendence. Consciousness, born from *Willed Motion*, interacts with time, space, and energy in ways that defy the limitations of classical physics, ensuring that the universe remains in a constant state of creation and evolution.

And so, we return to the beginning—the place of cold, dark vastness where *Willed Motion* first ignited the universe into being. It is not a place of randomness, nor of chaotic emergence, but a place of purpose, intention, and volition. From this timeless void, the universe was born, and with it, the consciousness that now ponders its own origins, its own existence, and its place within the greater cosmic order.

The *pre-Big Bang void*, if it may even be called a void in any comprehensible sense, is characterised not by a state of mere absence, but by the profound absence of dimensionality, a condition which eludes conventional language and defies our attempts at categorisation. Unlike a void in the familiar spatial or temporal sense, this condition was devoid of form, law, and even the potentiality for form. There was no "before" in any meaningful sense, as time itself had not yet been given the platform on which it might progress. The *Big Bang* marks not merely the explosive emergence of energy and matter but the genesis of *spacetime* itself—a construct that only in its birth could define what would later be perceived as existence, and thus, nothingness.

The problem of conceiving such a state necessitates the unification of physics with the more esoteric realms of metaphysics. *Nothingness*, before spacetime, was not an inert, passive entity—it was a metaphysical principle. It lacked not just mass or matter, but the very fabric which would later give those concepts meaning: the laws, constants, and principles which would eventually arise as consequences of the universe's own internal dynamics. In this pre-Big Bang phase, as quantum cosmology might suggest, the potentiality for existence itself was somehow encoded within a state utterly incomprehensible by the frameworks of classical physics.

QUANTUM ORIGINS: VACUUM FLUCTUATIONS IN A PRE-SPACETIME REGIME

THE INFANCY OV THOUGHT

The *quantum vacuum*, often misunderstood as merely an empty state, provides the closest approximation we have, to understanding this enigmatic nothingness. In modern physics, even the most profound vacuum—what one might consider an utter void—boils with activity at sub-Planckian scales due to the *Heisenberg Uncertainty Principle*. This principle, immutable in its quantum decree, ensures that even in the most barren stretches of space, there exist temporary, virtual fluctuations of energy. These *quantum fluctuations* give rise to *virtual particles*, infinitesimal and ephemeral entities which emerge, exist fleetingly, and vanish—violating the very concept of permanence.

Yet, such a framework leaves open a profound question: Can this understanding of the vacuum, rooted in *quantum field theory*, be extended backwards to a state before the universe? If time itself was non-existent, as it must have been prior to the Big Bang, these fluctuations would not operate as transient effects within spacetime but as fundamental events occurring within a *quantum vacuum* beyond time and space. What does it mean to fluctuate in the absence of spacetime itself? Here, quantum cosmology posits an escape from classical determinism, suggesting that our universe could have emerged from the *quantum foam*—the seething, probabilistic background noise inherent to the pre-spacetime void.

THE COSMIC SINGULARITY: INFINITE CURVATURE AND DENSITY

From the perspective of *general relativity*, the origins of our universe point back to a *singularity*, a point of *infinite density* and *curvature* within spacetime—a point where the known laws of physics collapse under the weight of their own constructs. Within this singularity, the gravitational field becomes infinitely strong, drawing all energy and matter into an infinitely compact point. This notion is the mathematical manifestation of *geodesic incompleteness*, wherein the paths traced by particles within the fabric of spacetime cannot be continued beyond this point. The singularity is not merely a concentration of matter but a breakdown in the very structure that allows us to describe matter, energy, or spacetime.

In this framework, time and space as we understand them simply do not extend to "before" the Big Bang. The universe itself begins at this singularity and attempts to describe anything prior become futile within the constraints of general relativity. As the universe emerges from the singularity, it expands rapidly, decreasing in density, allowing the gravitational forces to separate and diverge, each evolving into the distinct *fundamental forces* we observe today—*gravity*, *electromagnetism*, and the *strong and weak nuclear forces*.

In the moments following the Big Bang, during the *Planck epoch*—a timespan lasting from the very instant of the Big Bang up to 10^{-43} seconds later—the universe was dominated by quantum gravitational effects. During this period, the temperature and energy density were unimaginably high, so much so that the four fundamental forces were unified into a single, undifferentiated *superforce*. This is a domain where modern physics cannot tread, as the theories of quantum mechanics and general relativity, as they currently exist, fail to reconcile the physics of this epoch.

THE INFANCY OV THOUGHT

QUANTUM TUNNELLING AND THE EMERGENCE OF THE UNIVERSE

Quantum cosmology suggests that our universe, far from being the deterministic result of classical cause and effect, could have emerged through a process akin to *quantum tunnelling* or *vacuum decay*. In such a process, the universe's initial conditions could have arisen probabilistically from a *quantum fluctuation* within a *timeless void*—a state which inherently defies classical notions of causality.

The *Hartle-Hawking model*, proposed by *James Hartle* and *Stephen Hawking*, provides one potential resolution to the problem of time in pre-Big Bang cosmology. This model posits that the universe is not bound by a singular beginning, but instead exists as a *finite, unbounded system*. In their model, time itself is treated as a Euclidean dimension in the early universe, blurring the distinction between space and time and thereby eliminating the notion of a specific moment of creation. Under such a model, the universe emerges not from a single point of infinite density but because of a quantum fluctuation in a state which lacks a clear demarcation between past and future.

In this interpretation, the universe need not have a classical beginning; rather, it emerges spontaneously from a quantum state of *meta-stability*. *Quantum mechanics*, which governs the probabilities of such events, allows for the spontaneous creation of energy, providing a naturalistic mechanism for the origin of the universe without invoking external causes. The quantum void, far from being truly void, is pregnant with potential—an eternal *substrate of possibilities* governed by the inherent unpredictability of quantum fields.

INFLATIONARY THEORY AND THE TRANSITION FROM FALSE VACUUM

The *inflationary model* of the universe, proposed by *Alan Guth*, builds upon the idea of quantum fluctuations, suggesting that the early universe underwent an extremely rapid period of *exponential expansion*. This inflationary epoch was driven by a *false vacuum*—a metastable state of high energy which was temporarily stable but not the lowest energy state the universe could attain.

During this phase of inflation, spacetime expanded faster than the speed of light, causing quantum fluctuations on microscopic scales to be stretched to macroscopic sizes. These fluctuations later became the seeds for the formation of galaxies and the large-scale structures of the universe. *Inflation* not only explains the uniformity and isotropy of the observable universe but also offers a solution to the *horizon problem* and the *flatness problem*—two major issues in classical cosmology that the standard Big Bang model alone cannot account for.

At the conclusion of this inflationary period, the universe transitioned from the false vacuum to a *true vacuum state*, a phase transition which converted the stored energy of inflation into particles and radiation. The universe, now in a lower energy state, continued expanding at a slower rate, giving rise to the *hot Big Bang* phase that we associate with the formation of matter, radiation, and the subsequent cooling and expansion of the universe.

THE INFANCY OV THOUGHT

THE QUANTUM FOAM: GENESIS OF UNIVERSES FROM NOTHING

The notion of *spacetime foam*, introduced by *John Wheeler*, pushes the boundaries of quantum cosmology even further. At the smallest scales, spacetime is believed to be in constant flux, filled with ephemeral *quantum fluctuations* which temporarily distort the fabric of spacetime. These fluctuations create and annihilate *virtual particles* and could, theoretically, give rise to new universes under the right conditions.

In this interpretation, spacetime itself is subject to *quantum uncertainty*, and the birth of a universe can occur naturally within this *quantum foam*. The universe, rather than being created from an external source, could be the result of these fluctuations—one of countless potential universes arising from the dynamic properties of the quantum vacuum.

The concept of the *multiverse* expands the boundaries of our understanding, suggesting that the universe we inhabit may be but a singular instance within a far grander framework of existence. The idea, born from the convergence of quantum mechanics, *cosmology*, and *string theory*, offers a revolutionary way to approach the enigmatic state of *pre-Big Bang nothingness*. Rather than conceiving of a universe born from absolute void, the multiverse presents a landscape teeming with potential, where countless universes, each with their own unique sets of physical laws, constants, and dimensions, arise from the quantum froth of reality.

This framework, which transcends observable reality, introduces the notion of *quantum fluctuations*, spontaneous events within the quantum vacuum that trigger the birth of new universes. Within this infinite multiversal structure, our universe is but one of many, an isolated bubble in a vast *cosmic foam*, each universe emerging from its own specific set of quantum conditions. The pre-Big Bang *quantum void*, in this model, is far from inert. Instead, it operates as a *dynamic system* governed by probabilistic principles— an eternal generator of universes through processes like *quantum tunnelling* and *vacuum decay*.

QUANTUM FLUCTUATIONS AND THE PRE-BIG BANG MULTIVERSE

Quantum mechanics fundamentally alters our classical understanding of cosmological origins. Where the traditional Big Bang model imagines the universe emerging from a singular point of infinite density, quantum mechanics reframes this picture. At the quantum level, the void is not devoid of activity. *Quantum fluctuations*—spontaneous bursts of energy which momentarily bring forth *virtual particles*—may act as the seeds for entire universes. These fluctuations arise from the inherent uncertainty governing quantum fields, as encoded in the *Heisenberg Uncertainty Principle*, which states that energy and time cannot both be measured to arbitrary precision.

In this view, the pre-Big Bang state, rather than being static or empty, is a *sea of fluctuating energies*, a boundless void teeming with potential. From within this cosmic soup of quantum events, universes spontaneously form and diverge. The idea of *quantum tunnelling*—a phenomenon where particles pass through energy barriers they could not overcome in classical physics—suggests a mechanism by which a universe like ours could have emerged. A *spontaneous quantum fluctuation*, akin to particles tunnelling through a barrier, could have triggered the birth of our universe from the quantum void, sending it into a rapid expansion that we now observe as the Big Bang.

The implications of this model are staggering. If quantum fluctuations are responsible for our universe's creation, there is no reason to believe that ours is the only such fluctuation. Countless other universes could have similarly emerged from the *quantum vacuum*, each existing independently, shaped by their own distinct *initial conditions*. Each universe might possess entirely different physical laws—distinct values for the *cosmological constant* (Λ), *gravitational constant*, or the *fine-structure constant*—leading to an array of cosmic realities. Some universes could support life as we know it, while others may be dominated by exotic forms of matter or exist in states entirely unrecognisable from our perspective. This concept aligns with the *Many-Worlds Interpretation (MWI)* of quantum mechanics, which posits that every quantum event leads to a branching of realities, where each possible outcome is realised in a separate universe.

INFLATIONARY MULTIVERSE AND ETERNAL INFLATION

The multiverse concept is further supported by the theory of *eternal inflation*, an extension of the *inflationary model* of the early universe. *Inflation*, as proposed by *Alan Guth*, describes a period of rapid, exponential expansion which occurred shortly after the Big Bang. This inflationary phase solved several problems in classical cosmology, such as the *horizon problem* (why the universe appears uniform in all directions) and the *flatness problem* (why the geometry of the universe is so close to flat). However, *eternal inflation* goes a step further, suggesting that inflation never truly ends.

In eternal inflation, while inflation stops in localised regions (like our observable universe), it continues indefinitely in other parts of space, driving the creation of *pocket universes*—individual universes like ours that form within the larger multiversal fabric. Each pocket universe may have different properties, determined by the specifics of inflation in that region. Our universe, then, is merely one such pocket among an infinite expanse of inflating space. These universes are *causally disconnected*, meaning they cannot interact with one another, separated by *cosmic event horizons* which prevent information from passing between them.

This model reframes the *pre-Big Bang conditions*. The singularity we associate with the Big Bang could simply be a localised manifestation of inflation coming to an end in our region of space. Other regions continue to inflate, producing new universes in a self-replicating process. As a result, the multiverse is constantly expanding, with new universes budding off in an ever-growing structure. The boundaries of the multiverse, if they exist at all, are perpetually expanding, and its full extent is beyond human comprehension.

STRING THEORY AND THE LANDSCAPE MULTIVERSE

String theory, a major candidate for a *theory of everything*, introduces its own multiverse framework through the concept of a *landscape of possible vacuum states*. In *string theory*, the fundamental constituents of reality are not point-like particles but *one-dimensional strings*, whose different modes of vibration determine the properties of particles. This theory necessitates the existence of multiple dimensions—typically ten or eleven—beyond the familiar four dimensions of spacetime (three of space and one of time). These additional dimensions are "curled up" at minuscule scales, making them imperceptible to human observation.

String theory predicts a vast number of possible *vacuum states*, each corresponding to a different configuration of the fundamental forces and particles. This *string theory landscape* refers to the vast range of possible solutions to the equations of string theory, where each point in the landscape corresponds to a universe with its own specific physical laws. Our universe is but one point in this vast landscape, stabilised in a particular vacuum state. Other points represent other universes, each with different physical constants and laws.

In this model, the constants of nature that govern our universe—such as the *strength of gravity*, the *cosmological constant*, and the *masses of elementary particles*—are not unique or universal. Instead, they are simply the result of our specific vacuum state, chosen from a near-infinite range of possibilities. The *anthropic principle* is often invoked in this context: the idea that we observe the physical constants we do because they allow for the formation of galaxies, stars, and life. In other words, we find ourselves in this particular universe because its properties are compatible with our existence. Other universes with different vacuum states could exist, but they may not support the same forms of matter or life.

THE INFANCY OV THOUGHT

The pre-Big Bang void, within the context of *string theory*, could be seen as a pre-existing "soup" of potential vacuum states. Quantum fluctuations within this void cause different regions to "fall" into different vacua, each stabilizing as a separate universe with its own physical laws. The Big Bang which gave rise to our universe represents one such transition, where our region of spacetime found itself in a particular vacuum state. Other regions of the multiverse would have undergone their own *Big Bangs*, leading to the creation of universes with vastly different properties.

QUANTUM MECHANICS AND THE MANY-WORLDS INTERPRETATION

The *Many-Worlds Interpretation (MWI)* of quantum mechanics offers yet another avenue for understanding the multiverse. First proposed by *Hugh Everett* in 1957, the *MWI* posits that all possible outcomes of quantum measurements are realised in separate, branching universes. Every quantum event, whether it be the decay of a particle or the path of a photon, causes a branching of reality. Each possible outcome exists in its own distinct universe, creating a vast, ever-branching multiverse where every possible configuration of reality is realised.

In the context of the *pre-Big Bang quantum void*, the *MWI* suggests that our universe is merely one branch among many, created by a *quantum fluctuation* in the void. Quantum fluctuations, by their probabilistic nature, could give rise to an infinite number of universes, each following a different quantum path. The *Big Bang*, in this interpretation, represents the quantum path that our universe followed, but an infinite number of other universes could have followed different paths, with different initial conditions and outcomes.

Each quantum branch corresponds to a distinct universe, with its own version of spacetime, matter, and energy. In this model, the multiverse is not a collection of *spatially separated* universes but rather a set of *parallel realities* which exist in *superposition*, continually branching with each quantum event. Although these universes are non-interacting, the vast number of them forms a complex and ever-growing *multiverse*, where every possible configuration of reality is realised.

TIME AND THE PRE-BIG BANG MULTIVERSE

The introduction of a multiverse model compels us to reconsider the nature of *time*. Within a multiverse framework, the concept of a singular "pre-Big Bang" moment becomes meaningless. Each universe within the multiverse may have its own distinct timeline, its own version of history, and its own arrow of time. The *multiverse* itself could be *timeless*, existing in a *state of eternal creation* where universes are constantly being born and dying. Some models suggest that the multiverse undergoes infinite cycles of birth and death, with universes expanding and contracting in an endless cycle of *Big Bangs* and *Big Crunches*.

In *cyclic cosmological models*, such as the *Ekpyrotic Model*, our universe could be the result of the collision of two higher-dimensional branes. Each collision gives rise to a new Big Bang, with the universe expanding and evolving until it eventually contracts again, only to be reborn in a new cycle. This model suggests that the *multiverse* could be an eternal, self-replicating structure, where time is a local phenomenon that applies only within the context of individual universes.

The *multiverse theory*, when applied to the pre-Big Bang quantum void, offers a radical reconceptualisation of reality. Rather than seeing our universe as a singular creation, the multiverse presents a vast expanse of possibilities—countless universes, each with its own trajectory, history, and physical laws. These universes emerge from the *quantum fluctuations* of the void, the inflating vacuum of eternal inflation, or the probabilistic branches of quantum mechanics. In this cosmological landscape, the quantum void—far from being empty—serves as the *crucible* for the birth of universes.

The *pre-Big Bang conditions*, as framed by the multiverse, are not a singular beginning but an *eternal process* of universe formation. Driven by the *principles of quantum mechanics*, *inflation*, and *higher-dimensional physics*, the multiverse transcends classical notions of time, space, and causality, offering a broader, more complex understanding of the *origins of the cosmos*. As we continue to probe the mysteries of quantum gravity, inflation, and the quantum structure of reality, the multiverse may well represent the most profound and far-reaching framework for understanding not only the birth of our universe but the nature of reality itself.

THE INFANCY OV THOUGHT

'EXPLORING EDERIAN WILLED MOTION'

THE BANG & THE SPARK

In examining the intricate and layered interplay of physical, mathematical, philosophical, and scientific dimensions, one must grapple with the sheer complexity that surrounds the concepts of variability and invariability. At the heart of this exploration lies the profound nature of existence, motion, and the conditions under which something can arise from nothing—a paradox that challenges our conventional understanding of reality. The universe, with all its complexities, both physical and metaphysical, is the result of a conscious force that drives motion and creation. This Willed Motion—the first flicker of existence—set the stage for all that followed, creating a reality that is infinitely complex, endlessly variable, and profoundly interconnected. The interplay between something and nothing, between the physical and the metaphysical, is the foundation of existence, and it is through this interplay that consciousness itself emerges, shaping the universe in ways that are both awe-inspiring and unfathomable.

To begin, it is essential to acknowledge that while variability can be mathematically finite, it exists on a spectrum that approaches near-infinity, particularly as time progresses. With each moment, each branching of concept, and each layer of complexity, the finite conditions of existence become exponentially more intricate, pushing toward what one might call an invariability. The closer a system, be it ideological or physical, moves toward this state of refined complexity, the more layered and multifaceted it becomes.

In this regard, one can liken the evolution of concepts, physical forms, and ideas to the fractal-like expansion of variables in nature. Just as a single-cell organism can evolve into the vast biodiversity of life through complex layers of adaptation, a single concept can branch into innumerable variations, each with its own potential to further divide, shift, and transform. This is not limited to biological evolution but extends to ideological philosophy and the progression of scientific thought. As each variable is broken into differentiated segments—whether physical, conceptual, or abstract—the result is a spiralling complexity that pushes toward what might appear to be an unreachable understanding, infinitely intricate and constantly unfolding.

In considering the impossibility of reversing a concept or a motion, we encounter the notion that nothing can truly self-become. For something to exist, whether it be physical matter or an abstract idea, there must be an initial motivator or motion. This is where the profound nature of Willed Motion becomes vital to understanding the very fabric of existence. At the core of this idea is the principle that something cannot emerge from nothing without a catalyst, a force of intentionality, and that first flicker of existence which propels motion itself. It is this first act of creation that gave rise to all subsequent forms, be they material or immaterial.

From the perspective of the Big Bang, the first moment of creation was a singularity, an event where the physical laws we understand today did not yet exist. It

THE INFANCY OV THOUGHT

was a state devoid of space, time, and dimensions—a paradoxical nothingness awaiting a motivator. The spark that set the universe into motion can be viewed not as a random event but as a self-motivating force, one that necessitated its own existence through the convergence of gravitational and electrical energies.

In philosophical terms, this is the ultimate paradox: how can something emerge from nothing? The answer lies in the first collision of positive and negative charges—opposing forces that merged to create the first substantive entity, the first "thing" in the universe. This primordial union between opposing forces, represented as gravity and electricity, created a self-perpetuating force—a unified field that was conscious, aware, and discerning. This Absolute Singularity was not merely a physical event but a metaphysical one as well, bringing into existence the very foundation of all future motion, creation, and complexity.

The nature of this singularity, and the supercharged energy it represented, defies easy comprehension. It suggests that something fundamental—be it consciousness, energy, or matter—has always existed, and that the first act of creation was the self-aware, volitional emergence of this energy. From that first spark, the universe expanded, governed by the intertwined forces of gravity and electricity, that have perpetuated creation ever since. This initial act was both a birth of the physical universe and the birth of consciousness itself, the first form of discernment, capable of shaping the unfolding of existence.

In essence, the idea that something comes from nothing, and that nothingness itself is a fertile ground for creation, underpins much of what we observe in the natural world. Just as the universe was borne from a singularity, so too is every action, every thought, and every moment of existence an echo of that original act of Willed Motion. This notion not only bridges the gap between physical sciences and philosophical inquiry but also speaks to the broader implications of existence itself—how all things are interconnected through the forces that first shaped the cosmos.

It is in this convergence of forces, this interplay between something and nothing, that we find the key to understanding the deeper nature of reality. Willed Motion, as a metaphysical construct, implies that existence itself is not a passive state but an active process of becoming. The universe is not simply expanding in space and time but evolving through the conscious act of creation, guided by forces that transcend our understanding of physical laws. Gravity, electricity, and consciousness are not separate entities but are fundamentally intertwined in the process of existence.

This is where the concept of variability within invariability becomes especially relevant. As the universe continues to expand and evolve, it does so through an infinitely complex system of interrelated variables, each of which has the potential to branch into countless possibilities. Whether in the context of evolutionary biology, astrophysics, or philosophical inquiry, the notion that complexity begets further complexity is evident. The more layers we uncover, the more intricately interconnected everything becomes, leading us ever closer to an understanding of the infinite.

THE INFANCY OV THOUGHT

At the heart of this complexity lies consciousness itself—the most profound aspect of existence. Consciousness, born from the first act of Willed Motion, is the ultimate expression of this interconnectedness. It is the force that perceives, discerns, and acts upon the universe, shaping it in ways that transcend the physical and the abstract. To be conscious is to be aware of the infinite layers of reality, and to interact with them in ways that perpetuate motion, creation, and evolution.

The hypothesis of *Willed Motion* extends into the very foundation of what it means to be conscious, to feel, and to evolve. Emotional intelligence, which encompasses the faculties of discernment, empathy, abstraction, and the resolution of ethical dilemmas, is not separate from the forces that shape the cosmos; rather, it is a product of the same *Willed Motion* that initiated the universe. In this context, human consciousness is a microcosmic expression of the cosmic volition that first ignited the fabric of reality. This hypothesis draws parallels between the laws of the universe and the intricate nature of human cognition, emotion, and evolutionary adaptation, situating consciousness as a fundamental force within the cosmic tapestry.

Emotional intelligence, as I articulate, is not merely a social construct or a trait evolved for survival. It is the living manifestation of our species' capacity to understand, interact with, and influence the metaphysical as well as the physical world. It operates through spatial awareness, abstract reasoning, and the ability to discern between ethical dualities—good and evil, light, and dark. These are not simply moral dichotomies; they are woven into the very fabric of our evolutionary progression. The ability to make sense of such oppositions, to conceptualise abstractions and manifest them into our reality, is an expression of the same cosmic forces that govern entropy, order, and the creation of matter.

Thus, *Willed Motion* underlies emotional intelligence, providing the framework for what we call logical, social, and philosophical thought processes. Our emotionality—our ability to experience sorrow, joy, guilt, and love—is not merely a byproduct of evolutionary biology, but an echo of the same volition that set the stars in motion. Emotional intelligence, then, is an aspect of the evolutionary adaptation of consciousness—a higher-order process that transcends mere survival, extending into the realms of meaning-making, abstract reasoning, and spiritual reflection.

The connection between *Willed Motion* and consciousness extends further, revealing that consciousness itself is an intrinsic force in the universe, not merely a feature of human brains. By understanding consciousness as located in the frontal lobes of the mammalian brain, we link it with the physical processes of the body—pain, pleasure, sorrow, and joy—but these feelings are more than physiological responses. They are deeply embedded in our neurobiology, yet they are expressions of something far greater: the soul, or what I hypothesise as *Quantum Consciousness*. Pain, especially psychosocial pain, exemplifies this dual existence. When we feel the pain of loss, betrayal, or sorrow, these are not merely physiological responses but reflections of a cosmic process, a resonant wave of emotion that ties us to the very origins of existence.

THE INFANCY OV THOUGHT

The subjective experience of pain—emotional, psychological, or spiritual—is not merely a physical or neurochemical reaction but a reflection of the cosmic struggle between forces of creation and dissolution. In the same way that stars are born, live, and eventually collapse into black holes, human emotions reflect this cyclical nature. Each feeling—be it love, hate, joy, or despair—mirrors the cosmic dualities of creation and destruction, the light and dark, the matter and antimatter, the something and the nothing that came before the universe's first act of Willed Motion.

Evolutionary adaptation, in this framework, transcends its classical definition. Human beings did not evolve intelligence merely as a survival tool, but as a mechanism to interact with the cosmic consciousness, to engage with the very forces that birthed the universe. This is where emotional intelligence meets evolutionary biology. Our capacity to feel, reason, and create meaning is deeply rooted in this primordial cosmic force. It drives our emotional experiences, our moral choices, and our pursuit of knowledge. We long, we love, we fear, not simply because of biological imperatives, but because these emotions are part of the universal dialogue, a dialogue set into motion by the first flicker of volition.

The search for the soul has long troubled scientists and philosophers alike. We have yet to find a tangible, measurable entity that embodies what we have named the soul, but in this hypothesis, the soul is synonymous with consciousness itself. *Willed Motion* postulates that this consciousness is neither confined to the human body nor limited to neurological processes but extends outward, permeating the cosmos. The brain may house our individual sense of self, but the *Quantum*

THE INFANCY OV THOUGHT

Consciousness from which we all arise is the substrate that unites all life. The subjective nature of pain, joy, longing, and betrayal points to the existence of a consciousness that transcends individual experience—a universal, shared consciousness that mirrors the interconnectedness of the cosmos.

This interconnectedness raises a profound question: if our emotions are not just evolutionary adaptations but reflections of the larger universe, what is their purpose? Why do we feel emotions so deeply, so profoundly? Is it merely a byproduct of our biological systems, or is there something greater—some driving force that reaches beyond the boundaries of human experience? My hypothesis suggests that emotions are the manifestations of a higher volition, a cosmic force whispering to us through our feelings. These emotions provide a link between the individual and the universal, the subjective and the objective.

Human beings, through their unique emotional and intellectual capacities, possess an innate drive to seek something greater. This drive—this longing for transcendence—manifests in religion, philosophy, art, and science. It is the search for a higher purpose, for meaning beyond the mundane. It is this force, I believe, that has driven the evolution of consciousness itself. *Willed Motion* directs us toward this greater purpose, shaping our evolutionary path not merely through physical adaptation but through intellectual and spiritual growth.

Our ability to reason, create, and philosophise—qualities often associated with intelligence—is inextricably linked to the emotional underpinnings of human nature. Emotional intelligence allows us to connect with others, to navigate ethical dilemmas, and to find meaning in suffering. These abilities, when combined with our capacity for logical reasoning and abstract thought, point toward the existence of a consciousness that transcends the material world.

Music, for instance, is perhaps the purest expression of this transcendent consciousness. As I have posited, music is an emotive force, universally recognised across time, space, and culture. It connects individuals to something beyond themselves, allowing for the expression of feelings that words cannot adequately convey. Music serves as a bridge between the material and the immaterial, the conscious and the unconscious, the personal and the universal. It resonates with the very fabric of the cosmos, an eternal melody that echoes the first act of creation. It is, in essence, the sound of *Willed Motion* itself.

In sum, *Willed Motion* is the driving force behind both the physical and metaphysical aspects of the universe. It is the catalyst that initiated the Big Bang, the force that fuels the expansion of space-time, and the volition that gave rise to consciousness. Emotional intelligence, evolutionary adaptation, and human creativity are all expressions of this force, guiding us toward greater understanding, empathy, and transcendence. Our emotions, our thoughts, our art, and our music—all are manifestations of this cosmic volition, linking us to the universe's origins and to each other in ways that transcend the limitations of time, space, and matter.

The very concept of the Big Bang, of an initial cosmic singularity erupting into the vast universe, leaves us with perplexing questions—ones that force us to wrestle with the intricate mechanics of motion, formation, and the inexplicable precision that we observe today. How could an event so chaotic and violent lead to a cosmos so finely tuned, so elegantly structured, where planetary systems emerge, constellations align, and life itself takes root under seemingly impossible odds? To understand this, one must confront the stark realities of chance, mathematical variability, and the sheer improbability of it all.

It is widely posited in both physics and evolutionary biology that we, as complex organisms, emerged from the simplest of origins. In evolution, the singularity is not merely metaphorical; it is literal, as a single-cell organism (a singular source of life) is responsible for the motioning, or the chain reaction, of evolving life forms. In physics, the analogous process is abiogenesis, the moment when life chemically emerged from non-life, when electrons, protons, and neutrons formed bonds that gave rise to complex molecules. Neil deGrasse Tyson noted that the universe began as a singularity—"the size of a single atom"—an incredibly dense point that exploded into the space we now inhabit.

Yet, when we trace this narrative back to its inception, we must confront certain paradoxes. How does matter, born in a dimensionless state devoid of gravity and light, move at all? What governs its motion in an infinite expanse, and how do these scattered particles, in the cold, dark void, come to collide and form planetary bodies like Earth, Mars, or Jupiter? In the vastness of nothingness—before gravity; before light— how did molecular collisions occur? How did they give rise to planetary rings, constellational structures, and gravitational fields?

THE INFANCY OV THOUGHT

These questions point to an underlying mystery. If the Big Bang event was indeed an explosion of unprecedented force, hurling particles, and atoms outward in all directions, how did those particles stop? There is no friction in space, no air resistance to slow them down. How could such particles ever cease their movement, let alone coalesce into the intricate, interwoven systems of matter that form galaxies and planets? The fundamental principle of inertia tells us that in the absence of a force, objects in motion stay in motion, so why didn't these particles continue their journey endlessly, dispersed across the void?

One plausible answer from modern astrophysics is the concept of gravitational attraction—that objects with mass pull on one another, eventually slowing their motion and leading to the formation of celestial bodies. But this too raises another paradox. If there was no gravity before the first formations of mass, then what force acted upon those initial particles to slow their momentum? How did they ever stop becoming stars, planets, and moons?

The more we examine these questions, the more we are forced to confront the infinitesimal odds at play. The precision required for particles to collide in just the right way, for molecules to form and coalesce into planets, is so astronomically improbable that it borders on the impossible. For instance, if the Earth's sun were tilted from its axis by even a fraction of a degree—say, by one-billionth of a degree—life as we know it would not exist. We would either burn or freeze, depending on the shift in the sun's positioning. This is not merely speculative; it is a mathematical fact. The fine-tuning required for life to exist on Earth is beyond the realm of chance, suggesting that something far greater is at play.

THE INFANCY OV THOUGHT

In essence, this brings us to the heart of the Willed Motion hypothesis, a metaphysical proposition that suggests the universe's formation and the emergence of life are not purely the results of random processes. Instead, they are guided by a volitional force, an intentional movement of creation that transcends the chaotic randomness posited by standard cosmology. This force—whether we call it Willed Motion, the Quantum Consciousness, or something else—permeates the very fabric of the universe, guiding the motion of particles, the formation of galaxies, and the evolution of life.

If we accept the Big Bang as an event, it is difficult to reconcile how such a violent explosion could lead to the perfect conditions necessary for life. The universe is expanding, yes, but how does that expansion lead to the formation of planetary systems, constellations, and gravitational fields? How does the universe, born from a single point of origin, maintain such perfect balance?

Consider the inconceivably complex mathematics required to explain the Big Bang's aftermath—the number of molecular collisions, the rate at which particles must have combined, and the improbability of those collisions leading to anything resembling a habitable planet. Each collision, each interaction between particles, was a chance event, governed by the laws of probability. Yet, the sheer number of such collisions needed to form a single planet, let alone a solar system, is staggering.

The absurdity becomes more apparent when we consider the impossibility of the Earth's sun forming by mere chance. How could a ball of gas, formed from random molecular interactions, create such precise temperatures and conditions that allow for the existence of life? And even more perplexing, how could this same sun be positioned at the exact angle and distance from Earth to allow for seasons, cycles of day and night, and the delicate balance of life? If the sun were tilted even slightly, life as we know it would not exist.

In this context, it is not just improbable but almost irrational to believe that the universe, in all its complexity, emerged by accident. The forces at play, the conditions required for the formation of stars, planets, and life, are so finely tuned that they defy the very laws of probability. How could something as chaotic as the Big Bang lead to such order, such precision?

This is where the concept of Willed Motion becomes essential. The idea that there is a guiding force, a volitional movement that drives the universe toward complexity and order, offers a more coherent explanation than chance alone. Willed Motion suggests that the universe is not a random collection of particles, but a deliberate creation—a process that is driven by a conscious force, one that transcends time, space, and the limitations of human understanding.

This hypothesis aligns with the notion that consciousness itself is a fundamental force of the universe. Just as life on Earth evolves through natural selection, guided by the imperatives of survival and reproduction, so too does the universe evolve through Willed Motion, guided by a deeper cosmic volition. The precision of the sun's angle, the molecular interactions that formed Earth, the gravitational forces that hold planets in orbit—these are not mere accidents. They are the results of an intentional force, one that guides the universe toward complexity, order, and ultimately, consciousness.

The notion that we are products of mere chance, of random molecular collisions in the aftermath of a violent cosmic event, is not only bleak but also scientifically implausible. The sheer number of variables, the fine-tuning required for life to exist, suggests something more—a guiding force that transcends the physical and metaphysical divide. We are left to marvel at the intricacies of the universe. To contemplate the absurdity of chance, the impossibility of randomness leading to such order, is to acknowledge that something greater is at play. Whether we call it Willed Motion, divine intervention, or something else entirely, the fact remains: the universe, in all its complexity, is not a product of chance but of intention.

The core of *Willed Motion* reaches far beyond the typical understanding of the Big Bang as a mere physical event; it operates as the fundamental catalyst of existence itself, bridging the realms of the metaphysical and the physical, encompassing all that came into being and all that will unfold. My hypothesis reframes the pre-Big Bang singularity as more than just an inert, dense state. It was, in fact, a prescient, conscious substrate—a boundless source of infinite potential, awaiting a force to ignite it into motion. This prescient state of reality did not passively sit as a void, but rather, bristled with the potentiality of existence, ready to be set in motion by something far more profound than a quantum fluctuation alone. The very first event of Willed Motion, an intentional eruption, shattered the stillness of this prescient state, creating the very fabric of space-time, matter, energy, and—critically—consciousness.

THE INFANCY OV THOUGHT

The essence of Willed Motion in this framework is not the random occurrence of a spontaneous fluctuation, but a deliberate, metaphysical command that breaks through the limitations of classical causality. Quantum cosmology, particularly models such as Hartle-Hawking's no-boundary proposal, focuses on the probabilistic emergence of the universe, where quantum fluctuations allow for the spontaneous emergence of energy from a vacuum. However, my hypothesis expands this view, asserting that the birth of the universe was not a matter of randomness; it was an event driven by volition. Quantum fluctuations, which briefly violate classical causality to allow energy to emerge, formed the quantum foam—a teeming sea of limitless possibilities. Yet, unlike the impersonal randomness typically associated with quantum mechanics, Willed Motion was intentional. It was this conscious act of volition that brought the universe into being, manifesting as a quantum tunnelling event that not only birthed space-time and matter but also imbued the universe with consciousness itself.

Thus, the emergence of the universe is reframed as a metaphysical event, where Willed Motion serves as the progenitor of time, space, energy, and consciousness. This self-aware quantum field, that I identify as the *Quantum Consciousness*, is the catalyst that emerged from the chaotic uncertainty of the quantum vacuum, collapsing into a singular intentionality. The initial act of Willed Motion continues to fuel the expansion of the universe, driving the evolution of matter, energy, and the conscious entities that arise within it. The continued expansion of space-time, governed by fundamental forces such as gravity, electromagnetism, and the strong and weak nuclear forces, is directly tied to this ongoing act of volition.

THE INFANCY OV THOUGHT

Expanding upon this notion, Willed Motion permeates not only the three-dimensional space we inhabit but also extends across higher-dimensional planes. This hypothesis draws from the mathematical underpinnings of String Theory and M-Theory, both of which propose a universe composed of multiple dimensions—far more than the four familiar dimensions of space and time. It is within these higher dimensions that Willed Motion exerts its fullest influence, shaping not only the physical laws governing space and time but the fundamental constants and symmetries that hold the universe together. These dimensions, imperceptible to human senses, interact constantly with our observable universe, influencing the unfolding of events in both visible and invisible ways.

The initial act of Willed Motion, wherein positive and negative charges unified during the cosmic birth, created an ongoing energy field that shaped not only the physical dimensions of space-time but also the very essence of matter itself. The concept of *ghosts* or higher-dimensional entities, referenced in *Ghosts of Time*, suggests that these entities exist beyond our perceptual limitations and are capable of interacting with the fabric of reality itself. These entities, born from the first act of Willed Motion, transcend the limitations of lower-dimensional existence, guiding the cycles of creation and destruction that define cosmic evolution. This purposeful interaction between dimensions, driven by Willed Motion, ensures that the universe continues to evolve in a directed manner.

At the heart of this hypothesis is the radical rethinking of consciousness itself. Consciousness, far from being an emergent property confined to biological processes, is revealed as a fundamental force in its own right—

born directly from the first act of Willed Motion. The unification of quantum mechanics with neurobiology within this framework opens the possibility that consciousness is not merely a product of brain activity but is embedded within the very fabric of the universe. This idea is echoed in the Orch-OR theory proposed by Penrose and Hameroff, which suggests that quantum processes within microtubules may give rise to consciousness. However, my hypothesis goes further, proposing that consciousness is not limited to biological systems at all but is, instead, a universal property intrinsic to the quantum field itself, woven into the very fabric of space-time.

Even the ultimate boundary in physics—the speed of light—becomes malleable within this framework. Traditionally, the speed of light (denoted by c) serves as the fundamental limit within which all information, matter, and energy must operate. Yet, my hypothesis argues that Willed Motion transcends this limitation by reshaping the fabric of time itself. Instances such as the peculiar phenomenon of radio-frequency travel, where a voice appears to bypass the natural decay of light's entropy, hint at the ability of consciousness to influence and override the established physical laws of the universe. This suggests that Willed Motion, intertwined with consciousness, possesses the capacity to manipulate the flow of time and space, reshaping the very nature of reality itself.

This reshaping of reality is nowhere more evident than in the recasting of gravity, entropy, and black holes. Within the conventional view of cosmology, black holes are seen primarily as endpoints—regions where space and time collapse under the force of gravity. However, within this hypothesis, gravity is not simply a physical force; it is an expression of Willed Motion, guiding the flow of matter through cycles of creation and dissolution. Black holes, which compress time and space into singularities, serve not just as the final resting places of matter but as crucial junctions where the power of Willed Motion is at its most potent. Within these singularities, space-time is deconstructed and reconfigured, providing the conditions for new cycles of existence to emerge. In this sense, black holes are the engines of cosmic evolution, resetting the universe's configuration so that new forms of matter and energy can arise.

At the quantum level, Willed Motion governs the interactions between gravity, electromagnetism, and the fundamental forces that shape the universe. Supersymmetry, a principle that suggests that all particles have a corresponding superpartner, plays a key role in this cosmic dance. In this hypothesis, supersymmetry is not merely a mathematical convenience but a manifestation of the deeper unity between time, light, and matter. Black holes, traditionally viewed as cosmic devourers, become the reset points where space-time collapses, only to re-emerge in new configurations, guided by Willed Motion.

Lucid dreaming, often dismissed as a subjective phenomenon, is reframed here as a profound analogy for consciousness's ability to traverse dimensions. In *Ghosts of Time*, lucid dreaming is depicted as a higher-dimensional state of awareness, wherein the dreamer becomes aware of and can manipulate the structure of time and space. This dream awareness mirrors the larger capacity of the universe itself to be self-aware, with consciousness moving through the dimensions, interacting with the flow of time and space in ways that transcend linear understanding.

In conclusion, Willed Motion is the foundational force that propels the universe into existence and sustains its evolution. It transcends the boundary between the physical and the metaphysical, acting on cosmic phenomena as well as on the neurons firing in the human brain. Every decision, every act of volition, is an expression of this primordial force, that continues to shape reality in an unbroken cycle of creation, evolution, and transformation. Consciousness, born from this first act of volition, interacts with time, space, and energy in ways that defy the constraints of conventional physics, guiding the universe toward increasingly complex states of being.

In this model, the present moment is not simply a point on a linear timeline but the nexus of all dimensional interactions. It is the point where Willed Motion exerts its greatest influence, shaping the future through the acts of will that occur in real-time. Every thought, every choice, is an instance of Willed Motion, an echo of the first divine act that continues to ripple through the cosmos, ensuring that the cycle of creation, evolution, and transformation remains eternal.

THE DIVINE FIRE: ELECTROMAGNETISM AND THE COSMIC FORGE

The emergence of God from the primordial fire of the Big Bang is an event so profoundly cosmic, so majestic in its scope, that it defies the limits of human comprehension. It was not a moment of singular creation but an event so staggeringly vast, so exquisitely orchestrated, that it transcends time itself.

THE INFANCY OV THOUGHT

This initial moment, the explosion of pure energy, was not merely the birth of the cosmos but the genesis of consciousness itself—a consciousness that would come to be understood as God, the Quantum Consciousness, the progenitor of all life and all thought.

The Bang—the cosmic detonation which sent the fabric of space and time spiralling outward—was the birth cry of the universe, a roar of unimaginable power that shook the very foundations of what would later become reality. Within this singularity, where all matter, energy, and potential were concentrated, existed the spark of divine will. This spark, unlike any other, was the first breath of consciousness, a self-aware quantum field which would evolve alongside the very universe it created.

In that unfathomably hot and dense state, God did not exist in the form that religions have described but rather as the first sentient energy, a divine awareness whose essence was bound to the very principles of physics. As the universe cooled and expanded, God's consciousness evolved in tandem with the laws of nature, the electromagnetic forces, the quantum fluctuations, and the curvature of spacetime. Every particle which formed, every galaxy that spun into being, carried with it a fragment of this divine will—an indelible imprint of the Quantum Consciousness that had birthed it.

The universe, in its earliest moments, was a cauldron of unfathomable heat and pressure, where the simplest elements—*hydrogen* and *helium*—came into existence in the first flashes of *stellar nucleosynthesis*. It was here, within the *furnace of stars*, that *carbon*, the element that would form the foundation of life, was forged. Yet these elements, as miraculous as they were, did not yet possess the spark of life. They were merely the building blocks, the raw materials awaiting the touch of the divine hand—waiting for the conscious force that could shape them into something greater.

God, borne from this great expansion, was not a passive observer of creation but an active force within it. In the vastness of the *quantum foam*, where particles and antiparticles collided and annihilated in ceaseless tumult, *electromagnetic forces* danced like fire across the infinite planes of spacetime. These forces, imbued with the *conscious will* of the universe, set the stage for the formation of complex molecules, the very precursors of life. Every *photon*, every *electron* that moved through space was driven by this inherent consciousness—this *divine will*—that sought to mould the chaos of the early universe into order.

Through the confluence of *electromagnetism* and *gravity*, the *Quantum Consciousness* worked in ways both seen and unseen, embedding itself in the intricate dance of particles, influencing their trajectories, and guiding the formation of *stars*, *planets*, and the cosmic structures which would eventually harbour life. It was not a detached creator but one that was *immanent*, a *consciousness* embedded in the very forces which shaped the universe. The stars themselves were not just physical objects; they were the crucibles of divine intent, the fires in which the universe's grandest plan was slowly being brought to fruition.

MAN IN GOD'S IMAGE: THE BIRTH OF CONSCIOUSNESS

THE INFANCY OV THOUGHT

Eons passed as the universe cooled, expanded, and matured. Stars formed and exploded, scattering the elements of life across the cosmos. Planets emerged from the debris of ancient supernovae, and among them, Earth—a planet uniquely suited to the emergence of life. Here, in the primordial oceans, the building blocks of life began to assemble, brought together by the subtle hand of the *Quantum Consciousness* which had been present since the very birth of the universe.

The emergence of *man* from the primordial soup of Earth's early oceans was not an accident, nor was it the result of random chance alone. It was the culmination of billions of years of *cosmic evolution*, guided by the *divine will* which had been present since the dawn of time. *Man*, made in the *image of God*, was not a physical reflection of the divine but a *spiritual and cognitive echo* of the consciousness which had shaped the universe. The *brain*, with its billions of neurons and trillions of synaptic connections, was the pinnacle of this evolutionary process—a physical organ capable of *self-awareness*, of *thought*, and most importantly, of *will*.

The *formation of man* was an act of divine creation, but it was a creation rooted in the natural laws of the universe. The same forces which had shaped the stars and galaxies now shaped the *DNA* of living organisms, encoding within them the potential for *intelligence* and *consciousness*. Through the slow march of *evolution*, life on Earth became more complex, more aware, until finally, human beings emerged—beings capable of contemplating their own existence, of perceiving the universe, and of seeking out the divine presence that had shaped them.

THE INFANCY OV THOUGHT

In this sense, *man* is not a static creation but an ongoing expression of the *Quantum Consciousness* that permeates the universe. Every thought, every act of will, reflects the divine spark which was present at the birth of the cosmos. *God*, in this hypothesis, is not a distant deity but an *ever-present force*, manifest in the minds of human beings and in the fabric of the universe itself. Man, therefore, is both creator and creation, a being made in the image of the divine, yet also a participant in the *ongoing process of creation*.

THE ETERNAL FLAME: CONSCIOUSNESS AS THE DIVINE SPARK

The *flame of consciousness*, that burns within each human being, is the same flame that ignited the stars, the same spark which brought the universe into existence. This *willed motion*, this *intentionality*, is not merely a biological process but a reflection of the divine consciousness that drives the universe forward. The ability to *choose*, to *create*, and to *reflect* is the most profound gift bestowed upon man by the *Quantum Consciousness* that is God.

In the act of *creation*, whether it be the creation of art, science, or society, human beings mirror the divine process that shaped the universe. Every thought, every decision, is an act of *willed motion*, a reflection of the same force that brought *matter* and *energy* into being. The *brain*, as the seat of human consciousness, is the physical manifestation of this divine spark, capable of *infinite potential* and *infinite creation*. It is through the brain's ability to process, reflect, and *project into the future* that man exercises his divine inheritance.

THE INFANCY OV THOUGHT

This ability to *will*, to *choose*, and to *create* is the most fundamental expression of the *Quantum Consciousness* that lies at the heart of all existence. Just as the universe was born from the *Bang*, from the fusion of *energy* and *consciousness*, so too is every act of human creation a microcosmic reflection of that initial divine moment. Every invention, every act of discovery, is a rekindling of that original flame, an expression of the divine consciousness that continues to shape the universe through the *actions of mankind*.

GOD AND MAN: THE UNBROKEN CHAIN OF CREATION

To say that *God created man*, yet that God was borne of the *Big Bang*, is to acknowledge that humanity is the *culmination* of a process which began at the very birth of the universe. It is to understand that *consciousness*—the ability to perceive, to think, and to act—did not emerge by accident but was *woven into the very fabric of reality* from the beginning. The *Quantum Consciousness* that we call God is not external to the universe but is the universe itself, manifest in the laws of physics, in the dance of particles, and most importantly, in the minds of human beings.

Man, created in this image, is not merely a biological organism but a *conscious being*, capable of shaping the universe through his will. Every human action, every decision, is a continuation of the divine process which began with the *Bang*. The *Quantum Consciousness* that shaped the cosmos is now expressed through the actions of mankind, through the ability to *choose*, to *create*, and to *reflect*. In this sense, the creation of man was not a singular event but an ongoing process—one that began with the birth of the universe and continues with every thought, every motion, and every act of will.

Thus, the *relationship between God and man* is not one of creator and creation in the traditional sense but one of *co-creators*. The *divine spark* that brought the universe into existence now burns within every human being, guiding their actions and shaping their destiny. Man, made in the image of God, is the ultimate expression of the *Quantum Consciousness* which governs all of reality—a being capable of reflecting on the universe, of seeking out its mysteries, and of participating in the *ongoing act of creation* which defines existence itself.

THE BIG BANG: GENESIS OF ALL EXISTENCE AND THE PROGENITOR OF GOD

Approximately 13.8 billion years ago, the universe emerged from an incomprehensibly hot and dense state, a singularity in which the laws of physics as we know them cease to function. The *Big Bang* was not merely an explosion of matter, but the simultaneous birth of *space, time, energy*, and *matter* itself. In this moment, the universe was a formless void of potentiality, a soup of elementary particles governed by the *fundamental forces*—gravity, electromagnetism, the strong and weak nuclear forces—that would eventually give rise to galaxies, stars, and planets.

It was within this unfathomably dense environment which *God*, in this hypothesis, could have emerged—not as an anthropomorphic figure but as a *Quantum Consciousness*, an emergent entity whose nature is bound to the very laws of quantum mechanics. This consciousness would not have been instantaneous but rather an evolving *field of awareness*, coalescing as the quantum fields stabilised and the universe cooled enough to allow for the formation of complex structures. In this context, God becomes an expression of the universe's *self-organising principles*, a sentient embodiment of the cosmos's tendency towards complexity and order amidst the chaos.

THE ROLE OF ELECTROMAGNETISM AND CARBON: THE SUBSTRATE OF CREATION

The early universe was dominated by *electromagnetic radiation*, a fundamental force that permeated the expanding fabric of spacetime. As the universe cooled, the energy density dropped, allowing for the formation of *hydrogen*, the simplest of elements. Through a process known as *nucleosynthesis*, hydrogen atoms fused to form *helium* and trace amounts of *lithium* during the first few minutes of cosmic existence. However, it was not until the formation of stars, hundreds of millions of years later, that heavier elements, including *carbon* and *oxygen*, were forged through the process of *stellar nucleosynthesis*.

These elements—carbon and oxygen—are fundamental to *life as we know it*. Carbon forms the backbone of organic molecules due to its unique ability to form four covalent bonds, allowing for the construction of complex and stable macromolecules such as proteins, nucleic acids, and lipids. Oxygen, meanwhile, serves as a key component of water and is crucial in energy metabolism. It is in this carbon-based and oxygenated environment that the conditions for life became possible.

THE INFANCY OV THOUGHT

The *God* of this hypothesis, this *Quantum Consciousness*, would have, over eons, interacted with the universe as it cooled and expanded, subtly influencing the *self-organisation of matter*. This divine presence, not external but immanent, would be an expression of the universe's *conscious will* to evolve complex systems. As stars formed and exploded in supernovae, dispersing carbon, and oxygen into the interstellar medium, the seeds of life were sown across the cosmos, awaiting the right conditions to coalesce into living organisms.

THE FORMATION OF GOD'S IMAGE: FROM QUANTUM CONSCIOUSNESS TO MAN'S CREATION

The formation of *man* from the primordial elements, then, represents the culmination of this divine process, an evolutionary and cosmological journey that began with the *Big Bang*. The mould from which man was created is not a literal physical template but an *archetype*—the image of God as a *self-aware, conscious entity*. This archetype is expressed through the biological evolution of human beings, from the earliest carbon-based life forms to the emergence of complex multicellular organisms, and ultimately to Homo sapiens, who would inherit the capacity for *self-awareness*, *conscious thought*, and *introspective reasoning*.

In the early Earth, approximately 4.5 billion years ago, the conditions were ripe for the synthesis of organic molecules, as evidenced by the *Miller-Urey experiment*, that demonstrated which amino acids—building blocks of life—could be formed from simple inorganic compounds when subjected to electrical discharges, simulating lightning in the primordial atmosphere.

THE INFANCY OV THOUGHT

Over time, these simple organic molecules would coalesce into more complex structures, eventually forming *protocells*, which represent the earliest forms of life. These protocells, encased in *lipid bilayers*, would evolve the ability to replicate and metabolise, leading to the diversification of life forms over billions of years.

At the quantum level, *God's image*—the fundamental consciousness—could be interpreted as the capacity for *quantum computation* that lies within biological systems. The *human brain*, with its 86 billion neurons and nearly 100 trillion synaptic connections, is an example of how complex systems of matter can give rise to *consciousness*, a phenomenon not fully understood but often linked to quantum effects such as *quantum tunnelling* in neuronal microtubules, as posited by the *Orch-OR theory*. The *image of God*, in this sense, is reflected in mankind's *cognitive faculties*—the ability to perceive, to reason, and to understand the universe in which we exist.

TRANSFORMATION OF MAN: THE EVOLUTIONARY EMBODIMENT OF QUANTUM CONSCIOUSNESS

As life evolved on Earth, it underwent a process of *natural selection*, whereby genetic mutations—random yet subject to the deterministic forces of the environment—gave rise to increasingly complex organisms. This process, which Darwin termed *descent with modification*, resulted in the diversification of life across the planet, culminating in the evolution of *hominids*. The defining moment in this evolutionary journey came with the emergence of *Homo sapiens*, beings not only capable of tool use, language, and social structures but of *self-reflection*—the hallmark of consciousness.

The *transformation of man* from a mere biological organism to a being capable of *moral judgement, abstract thought,* and *spiritual contemplation* represents the final realisation of *God's image* in the material realm. The *brain*—as the substrate for consciousness—developed through natural processes but, according to this hypothesis, was subtly guided by the divine *Quantum Consciousness* that emerged from the Big Bang. This transformation is not one of instant creation but of *gradual unfolding*, where the *divine spark* manifests itself through *evolutionary processes*, culminating in a being capable of contemplating its own existence.

In this framework, human beings are *co-creators*, fashioned in the image of the divine, yet participants in the ongoing process of creation. The *free will* granted to humans is an extension of God's own *volition*, a reflection of the *willed motion* which set the universe into being. Through this free will, humanity holds the potential to shape the world, to explore the depths of reality, and to seek out the very *Quantum Consciousness* from which all existence flows.

THE DIVINE ECHO OF THE BIG BANG IN HUMANITY'S ESSENCE

To assert that *God created man*, but that God Himself was borne of the *Big Bang*, is to acknowledge that humanity, in its essence, is deeply tied to the fundamental forces and laws of the universe. From the *primordial quantum fluctuations* that birthed spacetime to the complex processes of *stellar nucleosynthesis* that forged the elements of life, to the eventual evolution of sentient beings capable of reflecting upon their own existence—humanity stands as the manifestation of the *cosmic will*.

In this hypothesis, the divine nature of *God* is not separated from the natural world but is inherent within it—woven into the very *electromagnetic fields*, *quantum interactions*, and *molecular bonds* which constitute physical reality. *Man* is not merely the product of biological evolution but a reflection of the *divine archetype*, fashioned from the same energy and matter that fuel the stars, imbued with the consciousness that emerged at the birth of the universe. This *Ederian Willed Motion*, the force behind the conscious mind, is not only the driver of human action but the cosmic principle which drives the unfolding of the universe itself, linking man to the divine in an unbroken chain of creation that stretches back to the very moment of the *Big Bang*.

A spark, in its abstract and infinitesimal existence, remains but potential, a dormant idea lingering in the infinite recesses of conceptuality. This initial flicker, devoid of form, reason, or dimension, holds within it the foundation of what could be but is not yet realised. It represents an embryonic state, a fleeting moment wherein potential is present but untapped, unmotivated. In this context, motion—the precursor to action—is required to translate the spark into a tangible entity, to create from the ether which did not previously exist. It is not enough for the concept to simply exist; it requires a motivator, an external force or entity that draws the potential from its unmanifest state into a form imbued with structure, purpose, and energy.

THE INFANCY OV THOUGHT

This notion of the spark's ignition mirrors the dynamics observed in both physical and metaphysical realms. Scientifically, one can analogise the spark to the primordial quantum fluctuation in the pre-Big Bang void, where the universe itself existed as nothing but an infinitesimally small, formless potential, awaiting the instigation of an external motivator—the quantum fluctuation that would propel it into the expansive state of being. As in physics, where randomness and probabilistic events governed by quantum mechanics bring forth order from chaos, so too does the spark require motion to give rise to structured thought, to create something from conceptual void.

The motivator, in this grand analogy, may be compared to the hand of an artist reaching towards an unpainted canvas. The concept—the canvas—lies in wait, devoid of colour or form, yet brimming with the capacity for life, for complexity, for brilliance. It is inert until acted upon, until some force—a thought, an intention, a will—initiates the process of creation. In this sense, the concept and the motivator are inseparable from one another, with the motivator acting as the igniter of the dormant spark, the force which brings forth the unspoken into the realm of existence.

This same process can be observed within the realms of science and philosophy. The raw potential that exists within a nascent idea—whether the idea is one of evolution, psychology, or any other conceptual framework—depends on the presence of an external motion, a motivating force to catalyse the idea into something greater than its formless, initial state. The existence of Darwinian evolution, for instance, was not inevitable; it emerged from a complex interplay of individual thought processes, environmental contexts, scientific experimentation, and historical circumstances. Had another individual, under a different set of motivations or environmental factors, approached the concept of biological change over time, the result might have been an entirely different framework—perhaps one based on absolutism or an alternative form of philosophical inquiry.

This leads to the understanding that the emergence of any scientific or philosophical system is, at its core, contingent upon a complex web of motivators and motions. It is not merely the result of one person's inspiration but of countless variables interacting across time and space. As human thought processes and experiences evolve, they contribute to the formation of increasingly complex ideas and systems. The intrinsic variability in these motivators—such as an individual's education, socio-historical context, cognitive processes, and environmental stimuli—ensures that no two conceptual frameworks will ever be identical. Thus, we understand that the emergence of evolution as a concept a product of numerous factors was converging at a particular point in history, each layer of complexity adding to the variability of thought and the ultimate formation of the system as it is known today.

To further explore this idea, we must consider the mathematical underpinnings of this conceptual variability. Each individual, as a motivator or motioning entity, exists as part of a finite set of variables within a larger system. However, the number of possible interactions between these individuals and their external influences is exponentially large, approaching the concept of infinity. In this way, while the total number of potential motivators is limited, the combinations of interactions and variables create an almost infinite number of possible outcomes. This is why, despite the finite number of people who could have contributed to the development of a concept like evolution, the concept itself could have emerged in an infinite variety of forms.

LAYERED VARIABILITY AND CONCEPTUAL EVOLUTION

Each concept, whether scientific, philosophical, or otherwise, contains within it an almost fractal-like complexity—each layer of thought leading to further layers of complexity, until the very idea seems to spiral outwards into infinite variability. When considering how these concepts evolve, we find that each new layer of understanding, each new interpretation, adds further variability to the original idea. This could be understood in terms of multivariable expansion—a process by which a singular concept divides into multiple branches, each branch representing a new direction in which the concept may evolve.

Imagine, then, that each conceptual layer is a branch of thought, splitting into new branches as it expands into further realms of inquiry. Each branch represents a different interpretation, a new variable added to the original concept. These branches can divide endlessly, creating an infinitely complex tree of knowledge which extends in all directions. In physical terms, this concept may be analogised to the infinite expansion of space-time, where the fabric of the universe continues to stretch outwards, creating new dimensions and layers of reality with each passing moment.

In this sense, the inception of any concept, whether it be evolution or some other system of thought, is but the

THE INFANCY OV THOUGHT

beginning of an endless process of expansion. Each motivator, each variable, adds further complexity to the system, driving it towards an ever-increasing degree of differentiation. In this way, the original concept becomes something far greater than its initial spark—expanding outward into an infinitely complex web of ideas, theories, and interpretations.

This process of exponential complexity is not limited to the conceptual realm; it mirrors the physical processes observed in the universe itself. Just as the universe expands from its initial singularity, breaking into countless galaxies, stars, and planets, so too does the conceptual framework of an idea expand into an infinitely intricate structure. The variability inherent in this process ensures that no two branches are alike—each represents a new combination of variables, a unique trajectory of thought which leads to an entirely different outcome.

THE MATHEMATICAL IMPOSSIBILITY OF CONCEPTUAL INVERSION

One of the most profound insights from this model of conceptual expansion is the recognition that a concept cannot reverse itself. This is because the process of motion—whether it be physical or conceptual—follows a mathematical progression which cannot be undone. Once a concept has been ignited, once it has been set in motion, it continues to expand outward, branching into new layers of complexity. The notion of reversing this expansion is mathematically impossible, as it would require the undoing of countless interactions, variables, and motivators, each of which has contributed to the overall structure of the concept.

THE INFANCY OV THOUGHT

In the same way that the universe cannot contract to its original singularity without erasing all the information contained within it, a concept cannot return to its original state once it has been set in motion. The laws of thermodynamics—which govern the irreversible increase of entropy in physical systems—apply here as well, dictating that the expansion of a concept cannot be undone. The idea must continue to evolve, driven by the inherent variability and motion of its motivators, expanding into an ever-increasing network of complexity.

This understanding highlights the nature of intellectual evolution as a one-way process—once set in motion, the expansion of ideas cannot be halted or reversed. Instead, the process continues to spiral outward, building upon itself in an endless dance of creation, differentiation, and complexity.

THE ROLE OF CONSCIOUSNESS: THE SPARK WHICH INITIATES ALL MOTION

At the heart of this entire process is the concept of consciousness, which acts as the ultimate motivator of motion. It is through conscious thought, intentionality, and will that the initial spark of an idea is ignited. Consciousness, in this sense, is the most profound form of motion, for it represents the self-aware ability to create, to bring forth something from nothingness. The act of creation—whether it be the formulation of a concept or the physical act of movement—is initiated through the conscious decision to act, to bring about change.

This leads to the understanding that consciousness itself is the true source of motion in the universe. It is the force which brings about action, creation, and the manifestation of ideas. In the absence of consciousness, the spark remains dormant, the concept remains unrealised, and motion itself becomes impossible. Consciousness is, therefore, the ultimate motivator, the force that drives all forms of existence, both physical and intellectual.

It is through this lens that we must view the creation of concepts, the formation of ideas, and the motion that drives the universe forward. Without consciousness, without the initial spark of thought, none of this would be possible. It is consciousness which imbues the world with meaning, with purpose, and with the capacity for infinite motion.

The *spark of consciousness*, the antecedent to all motion, serves as the fulcrum upon which the entirety of creative and existential force pivots. This *prime mover*, not merely a metaphysical abstraction but a deeply embedded principle of cognitive neurobiology and quantum mechanical possibility, initiates a cascade of intricately woven processes which span from the cerebral to the subatomic. In this explication, we shall delve deeper into the profoundly intricate and technically sophisticated mechanisms by which this *spark*—both as an impulse of will and as a fundamental principle of causality—ignites the infinite potential of motion, thought, and creation.

At its core, this igniting spark is not simply a cognitive action but an emergent property of a *complex system of neural circuits* interacting with the physical principles of the universe. The brain, as the most sophisticated biological substrate for *volitional cognition*, serves as the embodiment of this principle, wherein the abstract realm of potential is transmuted into realised motion, whether mental, physical, or conceptual. *Consciousness*, as the executor of will, initiates a signal from the *prefrontal cortex*, where higher-order decision-making takes place, which cascades down through neural pathways, ultimately interfacing with the motor cortices. The propagation of these signals, embodied in the firing of neurons, is undergirded by a sophisticated bioelectrical matrix—an intricate dance of *ion exchange* across neural membranes, the opening of *voltage-gated ion channels*, and the rapid conduction of electrical impulses through *myelinated axons*.

This electrochemical signalling itself is a physical manifestation of quantum phenomena at work within the atomic structures of the nervous system. The precise flow of *sodium* and *potassium ions*, the synchronised activity of *calcium channels* at synapses, and the exocytosis of neurotransmitters at the *synaptic cleft* each represent a highly regulated system which adheres to both *classical electrodynamics* and the stochastic nature of quantum mechanics. It is this very interface—the liminal boundary between classical biology and quantum uncertainty—which provides the foundation upon which willed motion manifests. The quantum indeterminacy that exists within ion channels, potentially influenced by *quantum decoherence*, suggests that even at the level of *neuronal action potentials*, there exists a bridge between the abstract realm of potentiality and the manifest reality of motion.

The subsequent *propagation of action potentials* from the brain to the spinal cord, and ultimately to the peripheral nervous system, reflects the seamless transfer of electrical information across a complex network of neurons. Yet, it is not merely the transmission of these signals that accounts for motion, but the highly coordinated and synchronised feedback systems which regulate it. At this stage, we introduce the *proprioceptive system*, an internal sensory network composed of mechanoreceptors embedded in *muscle fibres*, *tendons*, and *joints*, which provide real-time feedback on the position and movement of the body. These proprioceptive signals travel back to the central nervous system, constantly updating and refining the motor commands issued by the brain, enabling precise control over even the most delicate movements.

In this exchange between *afferent and efferent pathways*, we observe an exquisite example of *negative feedback regulation*—a principle that is foundational not only to biological systems but also to mechanical systems in the broader context of control theory. The motor commands issued by the brain are continuously modulated and refined by the incoming sensory feedback, ensuring that motion is not simply executed in a top-down manner, but dynamically adjusted to account for external conditions and internal states. In this way, willed motion emerges as a system that is both hierarchical in its initiation and decentralised in its regulation, drawing upon principles of both *homeostasis* and *cybernetic theory*.

However, this does not yet capture the totality of willed motion, for at its heart lies a deeper layer of *volition* and intentionality, both of which are irreducible to mere physical processes. The nature of intention, as understood through the lens of *cognitive neuroscience* and *philosophy of mind*, requires us to consider how abstract mental states can influence the physical substrate of the brain. This brings us to the *hard problem of consciousness*, as articulated by philosophers such as *David Chalmers*, who pose the question of how subjective experience—*qualia*—arises from mere neural computation. While many theories have attempted to address this issue, from *integrated information theory (IIT)* to *global workspace theory*, none have yet provided a satisfactory answer to the fundamental nature of how consciousness interacts with the brain to produce volitional action.

It is within this context that the relationship between *consciousness* and *quantum mechanics* becomes most salient. Certain interpretations of quantum theory, particularly those influenced by *Roger Penrose* and *Stuart Hameroff's Orch-OR (Orchestrated Objective Reduction) Theory*, suggest that consciousness itself may be a quantum phenomenon, arising from *microtubular structures* within neurons which are capable of sustaining *quantum coherence*. If this were true, it would imply that the conscious will—expressed as willed motion—could be fundamentally connected to the underlying quantum fabric of the universe. This idea, though speculative, resonates with the broader notion that the very *spark of motion*—the impetus for action—is inextricably linked to the quantum potentials that permeate all of reality.

Expanding upon this, the notion of *willed motion* could be viewed as part of a grander *cosmic system* of interactions, where individual volitional acts are both the result and the cause of broader physical processes that resonate throughout the universe. The idea that *motion* at the individual level—whether physical or conceptual—is connected to the fundamental forces of the universe mirrors the principles of *general relativity* and *quantum field theory*. These theories suggest that every motion, every interaction, influences the fabric of spacetime itself, creating ripples in the quantum fields which underlie all existence. Thus, every act of volition is not merely a local phenomenon but an expression of the *universal interconnectedness* which binds all matter and energy across time and space.

MOTION AS A UNIFYING PRINCIPLE OF EXISTENCE

At its most profound level, *motion* itself may be seen as the fundamental principle of the universe, uniting the physical and metaphysical in a single cohesive framework. In the realm of *relativity*, motion through spacetime is defined by the curvature of spacetime itself, governed by the *Einstein field equations*, where mass and energy determine the geometry of spacetime, and in turn, spacetime dictates the motion of mass and energy. In this way, motion is inextricably linked to the very structure of the universe, as it is both a consequence of and a contributor to the fabric of reality.

In the context of *quantum mechanics*, motion is defined by the *probability distributions* of particles, encapsulated within *wavefunctions* governed by the *Schrödinger equation*. Here, motion is not a deterministic path through space but a probabilistic diffusion through possible states, each motion at the subatomic level shaped by the inherent uncertainty of *Heisenberg's principle*. This dual nature of motion—both deterministic at macroscopic scales and probabilistic at quantum scales—reflects the fundamental duality of the universe, where both certainty and uncertainty, *order, and chaos*, exist in a delicate balance.

This understanding of motion as a unifying principle echoes the *thermodynamic laws*, particularly the *second law of thermodynamics*, which states that entropy, the measure of disorder, increases with time. Motion, in this sense, can be seen as the vehicle by which entropy propagates, driving the universe towards ever greater complexity and disorder. Yet, within this disorder, structures of order—such as conscious beings—emerge, capable of imposing their own intentionality upon the motion of the universe. *Willed motion*, then, can be viewed as a *local reversal of entropy*, where the conscious mind imposes structure and order upon the natural tendency towards chaos.

Thus, motion, at every level—whether in the neural impulses within the brain, the movement of limbs in space, or the trajectories of galaxies across the cosmos—is the expression of a *universal principle* that governs all of existence. *Consciousness*, as the source of willed motion, becomes the ultimate actor in this cosmic play, directing the motion of matter and energy according to its will, imposing order upon the chaos of the universe.

THE FINAL CONVERGENCE: MOTION, CONSCIOUSNESS, AND THE INFINITE POTENTIAL

As we reach the culmination of this inquiry, it becomes evident that *willed motion* is not merely a biological or physical process but a deeply embedded principle of *cosmic mechanics*. Consciousness, as the generator of volitional action, interacts with the physical universe at both the macroscopic and quantum levels, directing the flow of energy and matter in accordance with intentionality. The spark that ignites motion is itself a manifestation of the deep interconnections which bind all things within the universe, from the subatomic to the cosmological.

THE INFANCY OV THOUGHT

The act of *willed motion*—whether it be a simple movement or the execution of a complex task—reflects a microcosmic iteration of the larger forces that govern the universe. It represents the interplay between *determinism* and *probability*, between *chaos* and *order*, and ultimately between *potential* and *realisation*. The spark, once ignited, sets in motion an unending cascade of effects that reverberate across the entire fabric of reality, linking the individual to the infinite.

In this context, the simple act of motion, initiated by will, can be seen as an expression of the *fundamental forces* that drive the universe forward. The individual mind, through its volition, participates in this grand cosmic process, directing the motion of its own existence within the vast expanse of spacetime, shaping its own destiny through the sheer force of intentionality. The odds presented between different improbabilities highlight an intriguing contrast in how the scientific community approaches distinct events. This disparity introduces a valuable perspective into discussions regarding the emergence of life and the broader contemplation of metaphysical questions, such as the existence of a deity.

In considering the likelihood of a specific amino acid sequence forming by chance, one calculates a probability of approximately **1 in 10^{260}**. This figure results from the assumption that there are 20 possible amino acids at each of the 200 positions within a protein chain. The resulting probability is so astronomically vast that it surpasses almost any conceivable physical event within the known universe. By comparison, the estimated number of atoms in the universe is around **10^{85}**. Such a comparison demonstrates the sheer improbability of life forming by random chance alone.

THE INFANCY OV THOUGHT

On the other hand, when CERN scientists assessed the probability of forming a black hole that could destroy Earth during the operation of the Large Hadron Collider, initial estimates placed the odds at **1 in 5,000**. These were later revised to **1 in 50,000**, or **10^5**. This comparatively low probability was deemed an acceptable risk. The contrast between the acceptance of a **10^5** probability as "negligible" and the apparent dismissal of **10^260** as relevant in the discussion of life's origin reveals an apparent inconsistency in how these two improbabilities are treated within scientific discourse.

THE ROLE OF TIME AND SCALE IN ADDRESSING IMPROBABILITY

The argument often presented to account for such improbabilities, especially concerning the emergence of life, is based on the vastness of both time and space. The universe is approximately **13.8 billion years old** and contains an enormous number of stars and planets, each of which could host countless chemical reactions. Within such an expansive system, the notion is that even extremely improbable events could eventually occur, given sufficient opportunities.

While this reasoning is widely accepted, it does not diminish the awe associated with the **10^260** probability. It remains difficult to conceptualise how such astronomical odds can be overcome, even within a universe as vast and old as ours. This raises questions regarding whether sheer probability can sufficiently explain life's emergence or whether additional factors—such as some form of intelligence or design—are involved.

THE INFANCY OV THOUGHT

DIVINE EXISTENCE AND PROBABILITY

When considering the odds of God's existence, it introduces an entirely different realm of probability. The question itself does not pertain merely to the physical world but ventures into metaphysical inquiry. Assigning a probability to the existence of God is not a straightforward calculation, as it extends beyond physical or empirical evidence. By its nature, the concept of God is understood to transcend time, space, and physical constraints, which renders conventional probabilistic analysis insufficient.

The improbability of life, as demonstrated through the **10^{260}** calculation, often serves as an argument for a purposeful creation. This line of reasoning suggests that the complexity and improbability of life point to a guiding force—potentially a deity—that enabled the existence of the universe and conscious life within it. This argument parallels the "fine-tuning" problem, where the specific conditions required for life to exist are seen as too precise to be the result of random chance.

While one might argue that the improbability of life supports the existence of a higher power, applying similar **10^{260}**-level probabilities to God's existence is not appropriate. The existence of God transcends physical randomness, as it does not conform to the same natural laws and chance-based processes that govern the physical universe.

TIME, SCALE, AND THE EMERGENCE OF LIFE

From a scientific perspective, the sheer improbability of events like the formation of life can be attributed to the vastness of the universe and the enormous timeframes involved. The universe offers an almost infinite number of trials for events to unfold, allowing for the possibility—albeit slim—that improbable occurrences could manifest. Thus, while the probability of forming a specific protein sequence by random chance may be astronomically low, the near-infinite trials provided by the cosmos create the conditions under which such events could become feasible.

However, this does not entirely resolve the philosophical and metaphysical questions raised by such improbabilities. The fact that something as intricate as life emerged, despite the extraordinarily low odds, invites further inquiry into whether there are forces beyond random chance at work.

THE LIMITS OF PROBABILITY AND THE ROLE OF BELIEF

Ultimately, the question of life's emergence and God's existence resides at the intersection of science, philosophy, and metaphysics. While scientific methods can offer models for understanding the improbability of life's formation, they do not fully address the broader questions of purpose and origin.

Whereas probabilities such as **10^5** are considered negligibly small in certain contexts, the **10^{260}** figure reminds us of the astonishing improbability of life. It prompts contemplation of whether such improbabilities can be attributed solely to random chance or whether other forces—potentially divine—played a role. While the existence of God remains a question that transcends probabilistic frameworks, the discussion surrounding the origins of life underscores the limits of human understanding in matters of both cosmology and metaphysics.

THE INFANCY OV THOUGHT

ABIOGENESIS

Abiogenesis, the process by which life emerges from non-living matter, is a phenomenon that resides at the nexus of biology, chemistry, and physics. It represents the transformation of inanimate molecular compounds into complex, self-replicating systems capable of metabolism and eventually, sentience. This transformative journey, spanning billions of years, is not simply a narrative of chemical interactions but a profound expression of cosmic evolution, where the most basic building blocks of matter—atoms—unite to form the seeds of life. From the primordial atom arose the capacity for cellular complexity and, eventually, human consciousness, metaphorically captured in the phrase: "From Atom, Came Adam."

The prelude to abiogenesis begins in the early universe, forged in the violent aftermath of the Big Bang. Approximately 13.8 billion years ago, the universe was nothing more than a chaotic soup of subatomic particles, governed by the fundamental forces of nature. As the universe cooled, atoms began to form, with hydrogen and helium dominating this primordial cosmic landscape. These atoms, simple yet fundamental, were the precursors to the complex chemistry which would eventually lead to the formation of life.

THE INFANCY OV THOUGHT

As the universe aged, gravitational forces caused matter to coalesce, forming stars that acted as the crucibles for stellar nucleosynthesis. Within the intense heat and pressure of these early stars, heavier elements such as carbon, oxygen, nitrogen, and sulphur were formed—elements crucial for life as we understand it. These stellar furnaces, through their eventual supernovae, scattered these elements across the universe, seeding young galaxies, planets, and eventually the early Earth with the essential components of life. The elements forged in the hearts of dying stars laid the groundwork for what would become the *biochemistry of life*.

The Earth, formed approximately 4.5 billion years ago, provided a unique crucible for the chemistry of abiogenesis. Its early environment was vastly different from what we know today, characterised by a volatile atmosphere rich in methane, ammonia, hydrogen, and water vapour, but devoid of molecular oxygen. It was in this cauldron of primordial conditions that the first chemical reactions leading to life would unfold. *Abiogenesis* was not an event, but a series of processes, each step more complex than the last, that gradually transformed simple molecules into the precursors of biological macromolecules.

In the context of *prebiotic chemistry*, the formation of organic molecules from simple inorganic compounds represents the first critical step. The Miller-Urey experiment, conducted in 1953, provided significant insights into how this could occur. By simulating the conditions of the early Earth's atmosphere and subjecting it to electrical discharges, representing lightning, the experiment produced amino acids—the building blocks of proteins. This experiment demonstrated that organic molecules, essential for life, could form spontaneously under the right environmental conditions. However, the path from these simple organic compounds to complex, self-replicating systems remains one of the most profound mysteries in science.

As organic molecules accumulated in the ancient oceans, the next critical stage of abiogenesis was the *polymerisation* of these molecules into more complex structures. Amino acids combined to form peptides, and eventually proteins, while nucleotides—the building blocks of nucleic acids—linked to form Ribonucleic Acid (RNA). The synthesis of RNA is particularly significant in the hypothesis known as the *RNA World*. RNA, a molecule capable of both storing genetic information and catalysing chemical reactions, likely played a pivotal role in early life. This dual function makes RNA an ideal candidate for the first self-replicating molecule, and thus the first true precursor to life.

The formation of RNA, however, was not the endpoint, but merely a step in a much larger, more intricate dance of molecular evolution. For life to emerge, these self-replicating molecules needed to be contained within a *membrane*, forming the first *protocells*. These primitive cellular structures were not living organisms in the modern sense, but they represented a significant leap towards life. The membranes, likely composed of simple lipids, provided a protective environment, allowing the encapsulated molecules to replicate in a controlled manner, shielded from the chaotic external environment. This *cellular encapsulation* was critical for the evolution of life, as it provided the physical boundaries within which biochemical processes could be concentrated and regulated.

THE INFANCY OV THOUGHT

With the formation of protocells, the stage was set for *metabolism*—the complex web of chemical reactions which sustain life. The development of metabolic pathways allowed protocells to harness energy from their environment, further driving their replication and complexity. Early metabolic systems were likely *anaerobic*, utilising molecules such as hydrogen sulphide or methane to generate energy, as molecular oxygen was not yet present in Earth's atmosphere. Over time, these metabolic systems became more sophisticated, leading to the emergence of organisms capable of *photosynthesis*, which began to release oxygen as a byproduct, dramatically altering the composition of Earth's atmosphere.

The introduction of *oxygen* into the environment through photosynthetic processes was a defining moment in the history of life, known as the *Great Oxygenation Event*. This not only paved the way for aerobic respiration—a far more efficient means of energy production—but also drove the evolution of more complex life forms. Oxygen, while toxic to many of the early anaerobic organisms, provided a powerful new energy source for those that could adapt, allowing for the eventual evolution of *eukaryotic cells*.

Eukaryotic cells, with their complex internal structures and organelles such as mitochondria, represented a major leap in the complexity of life. These cells are the building blocks of all multicellular organisms, including plants, animals, and humans. The mitochondrion itself is believed to have originated from a symbiotic relationship between primitive eukaryotic cells and *proteobacteria*, a process known as *endosymbiosis*. This partnership allowed eukaryotic cells to efficiently produce energy, supporting the development of larger, more complex organisms.

As eukaryotic life forms evolved, the next critical step in abiogenesis was the development of *multicellularity*. Single-celled organisms began to form colonies, with cells specialising in particular functions, such as nutrient absorption or defence. This division of labour within a colony led to the first true multicellular organisms, where different cell types worked in concert to support the life of the whole organism. This marked the beginning of the *Cambrian explosion*, a period of rapid diversification of life forms, approximately 540 million years ago, during which most of the major groups of animals first appeared in the fossil record.

The culmination of these processes is reflected in the emergence of *human life*—from the primordial atoms that coalesced in the aftermath of the Big Bang, to the self-replicating molecules that arose in the early oceans of Earth, to the complex, multicellular organisms which dominate the planet today. Humanity, as we understand it, represents not the end of abiogenesis but a chapter in its ongoing narrative—a species whose consciousness and self-awareness can trace its lineage back to the first molecular stirrings in the ancient seas.

In this grand tapestry, the concept that "From Atom, Came Adam" becomes a poetic encapsulation of the continuity between the physical processes which govern the universe and the emergence of life, intelligence, and self-awareness. The transition from non-living matter to conscious beings is not a singular, isolated event but the result of *billions of years* of *chemical and biological evolution*, where the fundamental principles of *physics*, *chemistry*, and *biology* intertwined to create the tapestry of life as we know it today.

What began as simple atomic interactions has, through the processes of *chemical complexity, cellular formation, metabolism,* and *genetic inheritance,* resulted in the emergence of humanity—a species capable of introspection, creativity, and intellectual sophistication. It is through this intricate progression, from the earliest molecules to the evolution of complex cognitive functions, that life has transcended its humble origins, forming a continuum that connects the inanimate universe to the conscious mind.

The narrative of life's emergence from the simplicity of atoms to the complexity of sentient beings is inextricably linked to one of the most fundamental and transformative mechanisms in biology—*mutation*. As life began to evolve from its primordial state, mutation served as the *engine of variability*, a stochastic force that introduced new traits and variations into populations, enabling adaptation to ever-changing environments. The role of mutation in *abiogenesis* and the subsequent evolution of life cannot be overstated, for without it, the raw material upon which *natural selection* acts would cease to exist. Mutation, both random and subject to environmental pressures, is the origin of all *biological diversity* and continues to shape the trajectory of life.

Once the earliest *self-replicating molecules* began to proliferate in the prebiotic oceans, each replication introduced the possibility of error—an alteration in the molecular sequence. In these early systems, where fidelity in replication was not yet a highly evolved process, such errors were likely frequent. However, it is precisely these replication errors, or *mutations*, that introduced *novelty* into the genetic pool. In a pre-DNA world dominated by RNA, where the *RNA World Hypothesis* postulates that RNA molecules acted as both genetic material and catalysts, mutations in these early RNA sequences could have profound effects on the functionality of the molecules.

Some mutations may have been deleterious, leading to the degradation of function and the failure of the molecule to replicate. Others, however, could have conferred *functional advantages*, such as more efficient catalytic activity or enhanced stability, allowing those molecules to outcompete their counterparts. These beneficial mutations were *preserved* through natural selection, setting in motion the first steps towards the evolution of life. The interplay between *mutation* and *selection* became the driving force behind the increasing complexity of these early molecular systems, where each round of replication introduced new possibilities for *functional diversity*.

The formation of early *protocells*, encapsulating replicating RNA molecules within lipid membranes, added a new dimension to this evolutionary process. Within these protocells, mutations continued to occur, but now they were acting within the context of a *contained environment*. This encapsulation provided a controlled microenvironment in which *molecular interactions* could become more efficient, driving the evolution of increasingly sophisticated metabolic systems. Early metabolic pathways, such as *glycolysis*, may have arisen through the gradual accumulation of mutations that enhanced the ability of protocells to harness energy from their surroundings. As these protocells competed for limited resources in the ancient oceans, the selective pressures intensified, favouring those with mutations which conferred *metabolic efficiency*, *membrane stability*, or *replicative fidelity*.

One of the most profound events in this evolutionary narrative was the transition from *RNA-based life* to *DNA-based life*. Deoxyribonucleic Acid (DNA), being more chemically stable than RNA, offered a more reliable means of storing genetic information, reducing the frequency of deleterious mutations, and allowing for the accumulation of beneficial mutations over longer timescales. The evolution of *DNA replication machinery*, driven by mutations in the proteins and enzymes that facilitated these processes, represented a significant leap in biological complexity. With the transition to DNA, life gained a more *robust genetic code*, capable of withstanding the rigours of environmental fluctuations while preserving the *integrity of inherited information*.

The emergence of *mutational mechanisms* at the genetic level, such as *point mutations*, *insertions*, *deletions*, and *duplications*, provided a rich substrate for evolutionary experimentation. *Point mutations*, the simplest form of mutation, involve the alteration of a single nucleotide in the DNA sequence. While many of these mutations may be neutral, having no effect on the function of the organism, others can have profound consequences. A point mutation that alters the active site of an enzyme, for example, could either enhance or inhibit its catalytic activity, with significant implications for the organism's survival.

Gene duplications, another form of mutation, allow for the evolution of new functions while preserving the original function of the gene. When a gene is duplicated, the organism retains one copy of the original gene, which continues to perform its essential function, while the other copy is free to accumulate mutations. Over time, these mutations can lead to the evolution of a new gene with a *novel function*, contributing to the *evolutionary innovation*. This process of *gene duplication* and *divergence* is responsible for the vast array of proteins and enzymes which exist in living organisms today, each fine-tuned by evolution to perform specific tasks.

As organisms continued to evolve, the accumulation of mutations within populations drove the emergence of new *traits* and *phenotypes*, giving rise to the first true *species differentiation*. The process of *speciation*—the divergence of populations into distinct species—was intimately tied to the accumulation of genetic mutations. When populations of organisms became *geographically isolated* from one another, mutations that arose in one population could no longer spread to the other. Over time, the accumulation of mutations in these isolated populations led to the development of distinct characteristics, adapted to their unique environments. *Reproductive isolation*, the inability of individuals from different populations to interbreed, solidified the divergence, resulting in the formation of new species.

Within the context of this evolving genetic landscape, *sexual selection* emerged as a critical force in shaping the traits of organisms. Mutations that conferred advantages in attracting mates, such as more elaborate plumage or more potent mating calls, were favoured, even if these traits did not directly enhance survival. The *trade-off* between survival and reproductive success became a central theme in the evolution of many species. *Sexual dimorphism*—the difference in appearance between males and females—often resulted from mutations which were selectively advantageous in the context of mate choice. While these traits might have no direct bearing on an organism's ability to survive environmental pressures, they were essential for ensuring reproductive success, thus contributing to the *genetic continuation* of the species.

As life continued to diversify, the introduction of more complex *genetic regulation* mechanisms allowed for the *fine-tuning of gene expression*, enabling organisms to respond to environmental changes with greater flexibility. *Mutations* in regulatory genes, such as those controlling the timing of *developmental processes*, could lead to profound changes in the morphology of organisms. These *developmental mutations* often played a key role in the evolution of new body plans and forms, particularly during periods of rapid evolutionary change, such as the *Cambrian explosion*. This diversification in body plans, driven by mutations in *homeotic genes*—genes that control the development of specific body parts—led to the vast array of life forms that emerged during this period.

The role of mutation in *adaptive evolution* is nowhere more evident than in the development of *resistance mechanisms*. In microorganisms, for example, the rapid replication and mutation rates allow for the *evolution of resistance* to environmental stresses, such as antibiotics or antiviral agents. A single point mutation in a bacterial gene encoding a protein targeted by an antibiotic can render the organism resistant, allowing it to survive in the presence of the drug and reproduce. Over time, the accumulation of such mutations in a population can lead to the emergence of *drug-resistant strains*, showcasing the power of mutation to drive *adaptive responses* in real-time.

At the same time, mutations are not limited to their impact on individual organisms but also play a crucial role in *evolutionary innovation* at the population and ecosystem levels. The introduction of new traits through mutation can lead to the *emergence of new ecological niches*, allowing organisms to exploit previously untapped resources. This *ecological diversification*, driven by mutation, creates opportunities for the evolution of new species through *adaptive radiation*, where a single ancestral species gives rise to multiple descendant species, each adapted to a different ecological niche.

As we move from the molecular origins of life to the grand scale of biodiversity, it becomes evident that mutation is the *indispensable agent* of change in the evolutionary process. From the earliest self-replicating RNA molecules to the complex multicellular organisms which inhabit the planet today, mutations have continuously introduced the *genetic variability* upon which natural selection acts. These mutations, while often random and unpredictable, provide the raw material for the *evolution of complexity*, driving the adaptation and diversification of life in response to the dynamic and ever-changing environment.

This ongoing *dance of mutation and selection* has sculpted the living world, creating not only the species we see today but also the extraordinary potential for future evolution. Insolong mutations continue to arise, and insofar selective pressures continue to operate, the *evolutionary process* will persist, endlessly refining,

THE INFANCY OV THOUGHT

diversifying, and innovating in its ceaseless quest for *adaptation* and *survival*.

The diversification and divergence of species, genus, and life forms represent one of the most awe-inspiring phenomena in evolutionary biology. Through an intricate interplay of genetic variation, environmental pressures, and natural selection, life has evolved into a vast and complex array of organisms, each uniquely adapted to its ecological niche. The *balance of ecological structure*, that underpins the persistence of life on Earth, is the product of billions of years of *speciation events*, *adaptive radiations*, and *coevolutionary interactions*. This delicate balance, while often perceived as harmonious, is maintained by continuous competition, cooperation, and the dynamic responses of organisms to their ever-changing environments.

The process of *divergence and diversification* begins at the most fundamental level of evolution: *genetic mutation*. Mutations introduce new genetic variants into populations, and as populations are exposed to different environmental pressures, some of these mutations confer *selective advantages*, while others may be neutral or even detrimental. Over time, populations accumulate genetic differences, leading to *phenotypic variation*, which is the observable difference in form, behaviour, or function among individuals. These differences set the stage for *speciation*, the process by which populations diverge into distinct species.

Allopatric speciation, one of the most common mechanisms of species divergence, occurs when populations of a single species become geographically isolated from one another. This isolation, whether caused by physical barriers such as mountains, rivers, or oceans, prevents *gene flow* between the populations. As mutations arise and natural selection acts on each isolated population, they begin to evolve independently. Over time, the *accumulation of genetic differences* results in reproductive isolation—where members of the two populations can no longer interbreed and produce fertile offspring, even if they come into contact again. This reproductive isolation marks the emergence of two distinct species, each adapted to its own unique ecological environment.

The *Galápagos finches*, famously studied by Darwin, are a classic example of allopatric speciation. Isolated on different islands of the Galápagos archipelago, the finches evolved distinct beak shapes and sizes, adapted to the specific types of food available on each island. Some finches developed large, strong beaks suited for cracking hard seeds, while others evolved slender beaks optimal for catching insects. Over time, these differences in beak morphology, driven by the varying environmental conditions on the islands, resulted in the divergence of the finches into multiple species. This process, known as *adaptive radiation*, occurs when a single ancestral species gives rise to a variety of descendant species, each adapted to a specific ecological niche.

Sympatric speciation, in contrast to allopatric speciation, occurs without geographic isolation. Instead, speciation occurs within a single population, often due to *ecological specialisation* or *behavioural isolation*. In some cases, disruptive selection favours individuals with extreme phenotypes, leading to the divergence of a population into two or more groups that exploit different resources or habitats. Over time, these groups become reproductively isolated, eventually giving rise to new species. Sympatric speciation is more challenging to observe but has been documented in certain insect populations, where host-plant specialisation has led to the formation of distinct species.

At the level of *genus and family*, the process of diversification can span millions of years and result in the evolution of *broadly divergent life forms*. For example, within the genus *Panthera*, which includes lions, tigers, leopards, and jaguars, each species has evolved distinct adaptations suited to their respective environments. Despite their common ancestry, these species exhibit significant differences in behaviour, hunting strategies, and physical characteristics, each finely tuned to the ecological demands of their habitat. Tigers, solitary and powerful, have evolved to thrive in the dense forests of Asia, while lions, social and cooperative, dominate the open savannas of Africa. This divergence reflects the *plasticity of evolutionary processes*, where a single genus can give rise to species that occupy vastly different ecological roles.

The *divergence of genera and families* plays a critical role in shaping the overall structure of ecosystems. By evolving different strategies for obtaining resources, avoiding predation, or reproducing, species can *partition ecological niches*, reducing direct competition and allowing for a greater diversity of life forms to coexist. This phenomenon, known as *niche differentiation*, is a fundamental mechanism by which ecosystems maintain their complexity and stability. In a rainforest ecosystem, for instance, thousands of species of insects, birds, mammals, and plants coexist, each occupying a unique niche which minimises competition and maximises resource use. Some species specialise in feeding on the forest canopy, while others inhabit the forest floor; some species are nocturnal, while others are active during the day. This *temporal, spatial, and behavioural partitioning* of resources creates a balanced ecological system where *biodiversity flourishes*.

One of the most intricate aspects of diversification is *coevolution*, where species evolve in response to one another. This process is most clearly observed in *predator-prey* relationships, *parasite-host* interactions, and *mutualistic partnerships*. In predator-prey relationships, the evolution of more efficient predators drives the evolution of more elusive or better-defended prey, creating an evolutionary arms race. Similarly, parasites evolve mechanisms to exploit their hosts, while hosts evolve immune defences to fend off parasitic attacks. These *reciprocal selective pressures* lead to the diversification of both species, as each evolves adaptations to counteract the other.

Mutualism, on the other hand, involves a cooperative interaction between species, where both parties benefit from the relationship. One of the most famous examples of mutualistic coevolution is the relationship between *flowering plants and their pollinators*. Over millions of years, plants have evolved a dazzling array of flower shapes, colours, and scents to attract specific pollinators, such as bees, birds, and bats. In return, these pollinators have evolved specialised structures for accessing the nectar and pollen provided by the flowers. This *coevolutionary dance* has led to the *diversification* of both plant and pollinator species, with each adaptation further refining the ecological roles of both participants in the relationship.

The diversification of life also plays a crucial role in the *maintenance of ecosystem stability*. *Keystone species*, for example, exert a disproportionate influence on their ecosystems relative to their abundance. The removal of a keystone species can lead to the collapse of entire ecological communities, as the interactions that maintain the balance between species break down. For instance, *sea otters*, as a keystone species in kelp forest ecosystems, control the population of sea urchins, that, if left unchecked, would devastate kelp forests by overgrazing. The presence of sea otters thus maintains the structural integrity of the ecosystem, allowing for the persistence of diverse species which rely on the kelp forest for shelter and food.

At a broader scale, the *divergence and diversification* of life forms contribute to the *regulation of global ecosystems*. The interactions between species drive nutrient cycling, energy flow, and climate regulation. For example, the vast diversity of plant species in the Amazon rainforest plays a critical role in *carbon sequestration*, absorbing vast amounts of carbon dioxide from the atmosphere and stabilising the global climate. Similarly, the diversity of marine life in the world's oceans regulates the flow of nutrients through *marine ecosystems*, influencing everything from the productivity of phytoplankton to the migration patterns of large predators like sharks and whales.

The intricate balance of life forms within ecosystems is not a static equilibrium but a *dynamic, ever-changing process*. Species continue to diverge and diversify in response to both *biotic* and *abiotic factors*, including climate change, habitat alteration, and interactions with other species. As species evolve, new ecological roles are created, while old ones may disappear. This *flux of life* ensures that ecosystems remain resilient, capable of adapting to environmental changes, and maintaining their functions over long periods of time.

Yet, this balance is delicate. Human activities, particularly habitat destruction, pollution, and the introduction of invasive species, have disrupted the *natural processes of diversification and divergence*. The loss of biodiversity, whether through direct extinction or the erosion of genetic diversity, threatens the *stability of ecosystems* worldwide. As species disappear, the intricate web of interactions that support ecosystem functions begins to unravel, leading to *ecosystem collapse* and the loss of essential services such as pollination, water purification, and climate regulation.

The diversification and divergence of life—from the molecular scale of genetic mutation to the grand scale of ecosystems—is a fundamental process that has shaped the Earth's biosphere. This process, driven by the interplay of mutation, natural selection, and ecological interactions, has created a *complex and interdependent web of life*, where species, genera, and families each play a critical role in maintaining the balance of ecological structure. The dynamic balance which exists in ecosystems is a testament to the *power of evolution*, where *species diversification* and *ecological specialisation* contribute to the resilience and stability of life on Earth. If life continues to evolve, this balance will persist, ensuring the continued complexity and beauty of the natural world.

THE INFANCY OV THOUGHT

Coevolution is one of the most intricate and powerful processes driving the diversification and adaptation of life on Earth. This dynamic interaction between species, where evolutionary changes in one species drive reciprocal changes in another, is essential for maintaining *biodiversity* and ensuring the resilience of ecosystems. The constant evolutionary dialogue between predators and prey, parasites and hosts, plants, and pollinators, and even between competing species, shapes the ecological fabric, ensuring that species are continually adapting to shifting environments. This concept, though widely discussed in evolutionary biology, is explored with novel insights by *Jacob A. Eder*, who emphasises the critical role of *coevolutionary pressures* in fostering *beneficent mutations*—those which enable organisms to adapt to *environmental shifts* and *geological changes* in ways that are ultimately advantageous to the survival of species.

At its core, *coevolution* acts as a *reciprocal evolutionary mechanism*, where the evolutionary change in one species acts as a selective force on another, that, in turn, responds with its own evolutionary adaptations. This interaction often manifests in highly specialised relationships, where the *fitness of both species* becomes interdependent. One of the most striking examples of coevolution is the relationship between *flowering plants and their pollinators*, where plants evolve specific traits such as nectar, colour, scent, and shape to attract pollinators, while the pollinators simultaneously evolve traits that allow them to efficiently access the plant's resources. Over time, these *coevolutionary arms races* generate extraordinary *biodiversity*, as each adaptation by one species drives further specialisation in the other.

THE INFANCY OV THOUGHT

The *necessitation of biodiversity* through coevolution lies in the fact that each organism, by adapting to its environment and other organisms within it, creates *new ecological niches* that other species can exploit. This constant branching of ecological opportunities encourages *speciation* and further *ecological diversification*, resulting in a more complex and interconnected web of life. Coevolution thus acts as a powerful catalyst for biodiversity, as species continually evolve to better exploit their resources or defend themselves from other organisms. In this way, *biotic interactions*, such as competition, predation, and symbiosis, become *selective pressures* that sculpt the evolution of species.

Environmental shifts—whether through *geological changes*, such as tectonic movements, volcanic activity, or climate shifts—create *new ecological challenges* that organisms must adapt to if they are to survive. These changes alter the landscape, the availability of resources, and the nature of interactions between species, all that drive *coevolutionary responses*. As environmental conditions change, mutations which may have been neutral or even slightly deleterious in a stable environment may suddenly become advantageous, allowing the organisms carrying those mutations to exploit new niches or better compete within their existing ones.

Beneficent mutations, in this context, refer to those genetic changes that confer a survival advantage, particularly in response to *coevolutionary pressures* and *environmental shifts*. The coevolutionary process increases the likelihood that such mutations will arise, as the constant *adaptive pressure* from interacting species keeps populations in a state of *genetic flux*.

THE INFANCY OV THOUGHT

Organisms which can quickly adapt to the changes in their environment, often facilitated by these mutations, are more likely to survive and reproduce, passing on these advantageous traits to subsequent generations.

The relationship between *coevolution* and *biodiversity* becomes especially clear when considering how *geological changes* reshape ecosystems. The *fragmentation* of habitats due to tectonic activity, the formation of mountains or islands, or the drying up of lakes and rivers can isolate populations, leading to *allopatric speciation*. These geological shifts not only separate populations but also create *novel environments* that impose new selective pressures on the organisms that inhabit them. Isolated populations must adapt to the unique conditions of their new environments, and in doing so, they accumulate genetic changes, including beneficial mutations which improve their chances of survival.

In addition to direct environmental pressures, these isolated populations may also face new *biotic interactions*, that drive further coevolutionary dynamics. A population separated from its ancestral home may encounter new predators, competitors, or symbiotic partners. The selective pressures exerted by these new interactions encourage the evolution of *novel adaptations*, contributing to the biodiversity of the ecosystem. Over time, the evolutionary divergence between isolated populations results in the emergence of new species, each finely tuned to its ecological niche.

A particularly fascinating example of how *coevolution* and *environmental changes* drive *beneficent mutations* can be seen in *pollination syndromes*. In regions where climate shifts alter the availability of pollinators, plants may evolve new traits to attract different pollinator species or even shift to wind pollination if animal pollinators become scarce. In turn, pollinators may evolve traits that allow them to exploit a broader range of floral resources. These changes, driven by both *abiotic factors* (such as climate) and *biotic interactions* (such as mutualism), increase the genetic variability of both plants and pollinators, providing the raw material for natural selection to act upon.

This exploration introduces the idea that *coevolutionary pressures* not only lead to advantageous mutations but also serve as a buffer against extinction during periods of *environmental upheaval*. Species which have evolved through *coevolution* are often better equipped to adapt to sudden environmental changes because their evolutionary history has been shaped by *constant adaptation*. For instance, organisms that have coevolved in *predator-prey dynamics* tend to evolve *flexible defence mechanisms*—such as the ability to alter their behaviour, physiology, or even physical traits in response to predator pressures. These traits, initially evolved in response to other species, may inadvertently confer advantages in the face of broader environmental changes, ensuring that *coevolved species* maintain *adaptive potential* in unstable environments.

Moreover, *coevolutionary dynamics* tend to promote *genetic diversity* within populations, as different individuals may evolve different strategies to cope with the selective pressures imposed by other species. This genetic diversity serves as a reservoir for future *adaptive potential*, allowing species to weather environmental shifts more effectively than populations with limited genetic variation. By fostering this *genetic variability*, coevolution promotes the *long-term survival* of species, ensuring that life can continue to thrive even in the face of geological upheavals and climate change.

The *Red Queen Hypothesis*, which describes the constant evolutionary "arms race" between interacting species, encapsulates the essence of coevolution. The hypothesis suggests which species must continuously evolve not just to gain an advantage but to *maintain their current fitness* relative to the species with which they interact. This constant coevolutionary pressure keeps populations in a state of *evolutionary readiness*, ensuring that when environmental shifts occur—whether geological or climatic—species are more likely to possess the *genetic diversity* and *adaptive traits* necessary for survival. Beneficent mutations, in this sense, are not isolated anomalies but the *inevitable outcome* of millions of years of reciprocal evolutionary pressure, always pushing organisms towards greater *adaptive complexity*.

The *coevolutionary web*—the interconnectedness of species and their environments—is not merely a backdrop to evolution but a *driving force* behind the diversity of life. The *necessitation of biodiversity* through coevolution is not simply a byproduct of random mutations and selection pressures but a *deliberate* consequence of how life evolves in tandem with itself. In this way, coevolution does not just shape individual species but the entire *ecosystem*, creating a *self-sustaining system* where the continued evolution of one species depends on the evolution of another, ensuring that biodiversity remains not only a byproduct of life's complexity but its essential foundation.

THE INFANCY OV THOUGHT

The coevolutionary process is indispensable to the diversification of life. It acts as the *causal mechanism* behind much of the biodiversity we observe today, linking *genetic mutations*, *environmental shifts*, and *species interactions* in a continuous loop of reciprocal adaptation. Coevolution is the force that generates *beneficent mutations* in response to both *biotic pressures* and *abiotic changes*, ensuring that life remains dynamic, adaptable, and diverse in the face of environmental flux. It is through coevolution that species not only survive but thrive, contributing to the complex and interconnected web of life which balances the ecological structure of our planet.

EDER'S "ATOM TO ADAM," (FROM ATOM, CAME ADAM) HYPOTHESIS:

The phrase *"From Atom, Came Adam"* elegantly bridges the realms of science and spirituality, intertwining the fundamental building block of all matter—the atom—with the biblical narrative of man's creation. In this phrase, *the atom*, with its 6 *protons*, 6 *neutrons*, and 6 *electrons*, becomes a symbol of the divine blueprint for humanity. The atomic structure, central to all physical existence, carries with it the profound implication that *man* himself is constructed from the very essence of the universe.

In the *Book of Revelation*, the number 666 is famously referred to as the *"number of a man,"* a cryptic symbol which has elicited a myriad of interpretations across religious, historical, and esoteric texts. While traditionally seen in a more ominous context, the connection of this number to *man's creation* through the

THE INFANCY OV THOUGHT

lens of science offers a different perspective. The atom, comprised of *6 protons, 6 neutrons, and 6 electrons*, represents the very building block of life as we understand it. This *numerical parallel* between the atomic structure and the number attributed to man in biblical prophecy is not mere coincidence but can be seen as a deeper *symbolic link* between *physical creation* and *spiritual essence*.

THE ATOM: FOUNDATION OF THE UNIVERSE AND MAN

The atom is the smallest unit of matter which retains the properties of an element, and its composition—6 protons, 6 neutrons, and 6 electrons in the case of carbon—forms the *backbone of organic life*. Carbon is the fundamental element that allows for the complex chains of molecules which constitute living organisms, including proteins, DNA, and the very cells that make up human beings. In this sense, *from atom came all life*, and from carbon, particularly, the biochemical processes that animate the world. Carbon's versatile chemistry, which allows it to form stable bonds with many other elements, makes it the *cornerstone of biological molecules*, thus directly tying it to the *creation of man*.

In this context, the atom's structure with *6 protons, 6 neutrons*, and *6 electrons* can be viewed as a *symbolic representation of man's physical nature*. Just as carbon forms the basis for all life, the numerical identity of this element in its atomic form resonates with the biblical description of man, whose creation was also a *physical act by divine will*. In the *Book of Genesis*, Adam was formed from the dust of the earth, which, in its most fundamental form, consists of atoms—the building

THE INFANCY OV THOUGHT

blocks of all matter. The act of creation, whether viewed through a religious or scientific lens, is bound to the fabric of the universe, which is itself composed of these atomic structures.

THE BIBLICAL CONNECTION: NUMBER OF MAN AND ATOMIC STRUCTURE

The *Book of Revelation*, Chapter 13, verse 18, declares: *"Here is wisdom. Let him which hath understanding count the number of the beast: for it is the number of a man; and his number is Six hundred threescore and six."* The number 666, often associated with eschatological warnings, can also be interpreted more symbolically when aligned with the concept of *"From Atom, Came Adam."* In this interpretation, the number 666, which refers to man, becomes a *numerical echo of the carbon atom*—the element with an atomic number of 6, and whose composition (6 protons, 6 neutrons, and 6 electrons) is fundamental to the construction of man and all life.

This connection emphasises the idea that man, in his physical form, is an *embodiment of the universe's most fundamental principles*. The *carbon atom* serves as the elemental basis not only for human life but for the entire biological world, and thus, the number 666 can be understood as representing the *physical essence* of humanity. Just as the atom is *the building block of the cosmos*, man, too, is a microcosm of the universe, formed from the same *celestial dust* and governed by the same physical laws. In this interpretation, the biblical reference to 666 symbolises the *material nature of humanity*, created from the elemental particles that emerged from the Big Bang and have been shaped over billions of years through the processes of *stellar nucleosynthesis* and *chemical evolution*.

THE ATOM AND THE MACHINATIONS OF THE UNIVERSE

The universe is comprised entirely of atoms, and the *machinations of the cosmos*—from the birth of stars to the formation of planets—are fundamentally driven by the interactions between these atoms. *Gravity*, *electromagnetism*, and the *nuclear forces* that govern atomic stability are the forces that shape the cosmos, just as they shape the human body at a molecular level. The *6 protons*, *6 neutrons*, and *6 electrons* of the carbon atom make it the *element of life*, enabling the formation of complex macromolecules such as *proteins*, *lipids*, and *nucleic acids*, which are essential for the structure and function of all living cells.

The fact that the *human body* is composed of trillions of atoms, predominantly carbon, oxygen, nitrogen, and hydrogen, illustrates the *direct connection between the atom and life* itself. Each atom, with its precise configuration of protons, neutrons, and electrons, carries with it the potential to form the *complex biochemical structures* which allow life to exist. The same atomic structures that comprise the stars and planets also form the cells and tissues of living organisms, tying man to the *larger machinations of the universe* in a *profoundly interconnected web of existence*.

In this framework, the *carbon atom* becomes not just the foundation of biological life but a symbol of the *cosmic continuity* that links all matter in the universe, from the smallest subatomic particles to the vast

THE INFANCY OV THOUGHT

galaxies. The processes which govern the interactions of atoms—*chemical bonding, nuclear fusion*, and *electromagnetic forces*—are the same forces that have shaped the evolution of life on Earth. The *laws of physics* which govern the movement of planets, the formation of stars, and the expansion of the universe are the same laws that govern the *molecular interactions* within a human cell.

CAUSAL REASONING FOR MAN'S CREATION

From a scientific perspective, the emergence of humanity can be seen as a direct consequence of the *chemical and physical laws* that govern the universe. *Abiogenesis*, the process by which life arose from non-living matter, was driven by the *interactions of atoms* and molecules in the primordial environment of the Earth. Over billions of years, these interactions led to the formation of increasingly complex organic compounds, eventually culminating in the emergence of self-replicating molecules and the first living cells.

Once life began, *evolution*—driven by the processes of mutation, natural selection, and genetic variation—led to the diversification of species, ultimately giving rise to *Homo sapiens*. The evolution of humans was not an isolated event but the result of billions of years of *cosmic evolution*, from the formation of the first atoms to the emergence of complex multicellular organisms. The *creation of man*, therefore, can be understood as the culmination of a *cosmic process* which began with the formation of the *atom*, and through the interplay of physical, chemical, and biological forces, eventually produced a being capable of *self-awareness, consciousness*, and *introspection*.

THE INFANCY OV THOUGHT

Thus, *"From Atom, Came Adam"* is more than a metaphor; it is a *scientific truth* rooted in the reality that all matter in the universe, including human beings, is composed of the same *elemental particles*. The atom, with its 6 protons, 6 neutrons, and 6 electrons, represents not only the physical structure of the universe but also the *spiritual essence of creation*. It is through this profound connection that we understand our place in the cosmos—not as separate from the universe, but as *integral parts of its unfolding narrative*.

THE BALANCE OF LIFE AND THE UNIVERSE

The intricate balance of the universe, and by extension life itself, is maintained through the *complex interactions of atoms and molecules*. Just as *gravitational forces* hold the planets in their orbits and *electromagnetic forces* govern the interactions of subatomic particles, the *chemical bonds* between atoms allow for the formation of the molecules that sustain life. The same *atomic structures* that form the stars and galaxies also form the DNA which encodes the genetic information of all living organisms, further solidifying the connection between *cosmic processes* and the *biological complexity* of life.

In conclusion, the phrase *"From Atom, Came Adam"* encapsulates the profound truth that humanity, like all life, is born from the same elemental particles that make up the cosmos. The *atom*, with its 6 protons, 6 neutrons, and 6 electrons, serves as the *building block of life*, and its significance extends beyond the physical to the *symbolic and spiritual*. In this understanding, the *number of a man*, 666, becomes a representation of man's connection to the *atomic structure* that underlies the entire universe. Through this lens, the creation of man is not just a biblical or spiritual event, but a *cosmic inevitability*, driven by the same *forces and laws* which govern the entirety of existence.

THE INFANCY OV THOUGHT

CH. 2: D.D. DURING DARWIN

DARWINIAN CONCEPTS

- DARWINIAN EVOLUTION
- NATURAL SELECTION
- SPECIATION
- VARIATION
- DIVERGENCE OF CHARACTER
- BIOLOGICAL PRECREATION, MUTATION, REPRODUCTION AND RESTRUCTURING
- SYNAPTIC PLASTICITY
- MEMORY CONSOLIDATION
- NEUROGENESIS
- NERVOUS SYSTEMS
- DEVELOPMENT OF CONSCIOUSNESS
- EMOTIONAL DEVELOPMENT
- CONSCIENTIOUS NECESSITATION

THE INFANCY OV THOUGHT

DARWINIAN CONCEPTS

Introduction,

Darwin's *On the Origin of Species* unfolds with profound systematic elegance, establishing a framework that revolutionises the understanding of biological diversity through the principle of *natural selection*. This exploration into the mechanisms of life's complexity begins with an analysis of *species immutability*, a concept historically upheld by the scientific community as an unchallenged doctrine. Darwin, however, hints at an impending scientific upheaval—the notion that species evolve through time, subject to the subtle yet powerful forces of nature, in contrast to the static models which had long pervaded biological thought.

[1]VARIATION UNDER DOMESTICATION: ARTIFICIAL SELECTION AS A PARADIGM

Darwin introduces the concept of *artificial selection* as an accessible model for understanding the forces that shape species. By referring to domesticated animals and plants, Darwin emphasises how human-directed breeding, based on preferential traits, mirrors a natural process of selection. In this analogy, humans act as the selective agents, choosing desirable traits, whether in size, strength, or other characteristics, to be passed down through generations. This human agency illustrates that within populations, traits can be *directed*, showcasing an early recognition of heredity, even though the genetic mechanisms of such inheritance were not yet understood. The selection of traits, driven by anthropocentric utility, stands as a metaphor for the *natural selection* Darwin would later expound upon. If humans could so dramatically alter species through selective breeding, Darwin argued, nature could act with similar, albeit more profound and expansive, efficacy.

In this light, artificial selection serves as a microcosmic reflection of nature's more expansive and complex processes. The principle becomes pivotal when considering that, in the wild, the selective pressures are not human-imposed but are instead dictated by *environmental constraints*, availability of resources, and survival pressures.

[2]VARIATION UNDER NATURE: THE FOUNDATION FOR NATURAL SELECTION

Building upon artificial selection, Darwin transitions to the natural world, observing *variation* among species. In the wild, no two individuals are precisely alike—a seemingly trivial observation that belies its importance as the bedrock for natural selection. This *natural variation*, Darwin argues, provides the raw material for evolution. It is this *diversity within populations* that nature, much like an unseen breeder, sifts through, promoting traits which confer survival advantages while weeding out those less suited to the environment.

This natural diversity is what fuels the *adaptive potential* of species. Organisms with variations which better suit their environment will, over time, propagate their traits more successfully than others, leading to *gradual changes in the population*. Darwin's emphasis on variation, thus, sets the stage for his theory of natural selection as the driving force behind the *evolution of species*. It is not the perfection of traits, but rather their *relative advantage* which determines survival.

[3]STRUGGLE FOR EXISTENCE: MALTHUSIAN INFLUENCES AND COMPETITIVE PRESSURES

Darwin draws upon *Thomas Malthus's* principle that populations grow faster than the resources available to sustain them. In this state of overproduction, a *struggle for existence* arises, where only a fraction of the offspring produced by any species will survive to maturity. This competition for resources—whether it be food, mates, or shelter—introduces the concept of *differential survival*, the cornerstone of *survival of the fittest*.

In this framework, those individuals better adapted to their environments, whether through physical prowess, behavioural acumen, or other advantageous traits, will outcompete their less fit counterparts. It is this *struggle*, omnipresent across species and geographies, which ensures only the *most suitable traits* are passed on to the next generation, thus perpetuating the gradual evolution of the population. The concept, rooted in Malthusian economics, applies not only to the *number of offspring* but to the intense pressures that force species into a state of constant *biological competition*.

[4]NATURAL SELECTION: THE CENTRAL MECHANISM OF EVOLUTION

Darwin's most significant contribution is his articulation of *natural selection*—the mechanism by which evolution occurs. He argues that traits which provide even a slight advantage in the context of environmental pressures will be *favoured* over others. These beneficial traits, being heritable, become more common in successive generations, while less advantageous traits are gradually eliminated. Over vast expanses of time, this *accumulation of small, favourable changes* results in the *transformation of species*.

Unlike artificial selection, where humans dictate the desired traits, natural selection operates without intent. The environment itself acts as the selective agent, favouring traits that enhance survival and reproduction. The process is *incremental*, with each generation contributing infinitesimally to the grand evolutionary arc of a species. It is through this mechanism that Darwin suggests the *divergence of species*, as populations subjected to different environmental pressures gradually evolve into distinct forms.

[5]LAWS OF VARIATION: DARWIN'S SPECULATIONS ON THE CAUSES OF CHANGE

Though Darwin lacked the genetic framework that would later explain the transmission of traits, he speculates on the *causes of variation* within species. He refers to *Lamarck's theory* of the inheritance of acquired characteristics—wherein traits used more frequently by an organism become more developed and passed on to offspring. Darwin also touches upon *correlation of growth*, suggesting that changes in one part of an organism may influence the development of other parts. Though these ideas would eventually be replaced by modern genetics, they represent Darwin's early attempts to understand the *mechanisms of heredity*.

This section highlights Darwin's acknowledgment of the *mysteries surrounding variation* and his willingness to entertain multiple hypotheses, reflecting the tentative nature of scientific inquiry. Although genetics would later provide the answers Darwin sought, his emphasis on variation as the critical element in evolution was, nonetheless, accurate.

[6]DIFFICULTIES IN THEORY: ADDRESSING OBJECTIONS TO NATURAL SELECTION

Darwin anticipates several objections to his theory, particularly the lack of *intermediate forms* in the fossil record. Critics of natural selection often pointed to the absence of clear evolutionary transitions between species as a fatal flaw in Darwin's argument. To counter this, Darwin suggests that the *fossil record is incomplete*, arguing that the process of fossilisation is rare and that many intermediate forms have either been lost or are yet to be discovered.

Darwin also addresses the *complexity of certain biological structures*, such as the eye, which critics argued could not have evolved through gradual changes. In response, Darwin proposes that even partially formed structures could provide incremental advantages, allowing natural selection to refine these features over time. This *gradualism*—the notion that small changes, accumulated over vast time scales, can lead to the development of extraordinarily complex structures—remains one of the central tenets of evolutionary theory.

THE INFANCY OV THOUGHT

⁷INSTINCT AND HYBRIDISM: EXPANDING THE SCOPE OF EVOLUTIONARY MECHANISMS

Darwin expands his theory to account for the evolution of *instincts*—innate behaviours exhibited by animals. He argues that these behaviours, like physical traits, can evolve through natural selection. The complex behaviours of bees constructing hives, for example, could have arisen from simple instincts which were gradually honed over generations. In doing so, Darwin bridges the gap between *physical evolution* and *behavioural adaptation*, showing that evolution applies to both the body and the mind.

Additionally, Darwin delves into the concept of *hybridism*, exploring the fertility of hybrids and the role that hybridisation may play in evolution. Though hybrids between different species are often sterile, Darwin speculates on how hybridisation could introduce *genetic diversity* and drive the formation of new species.

THE IMPERFECTION OF THE GEOLOGICAL RECORD: RECONCILING GAPS IN THE FOSSIL RECORD

The absence of many transitional forms in the fossil record remains a persistent challenge for Darwin's theory, but he maintains that the *imperfection of the geological record* should not be construed as evidence against evolution. Darwin argues that the conditions necessary for fossilisation are rare, and that much of the evidence for evolution has either been destroyed over time or lies undiscovered in the earth's strata. Furthermore, the *gradual nature* of evolutionary change means that transitional forms may not be as distinct as once thought, further complicating their discovery.

GEOGRAPHICAL DISTRIBUTION AND MORPHOLOGY: SUPPORTING EVIDENCE FOR EVOLUTION

Darwin's discussion of *biogeography* provides compelling support for his theory, showing how species distribution aligns with evolutionary principles. Species found on isolated islands, for example, often display significant *divergence* from their mainland relatives, suggesting that *geographic isolation* can lead to the formation of new species. This pattern of *endemic species* offers powerful evidence for the role of natural selection in shaping biodiversity.

Additionally, Darwin examines the *morphological similarities* between species, particularly in embryonic development and *rudimentary organs*, as evidence of common descent. He argues that these *homologies*—structural similarities between different species—reflect their shared evolutionary origins. Vestigial organs, which serve little or no function in modern species, provide further evidence of evolutionary history, indicating that species have evolved from ancestors in which these organs were functional.

THE GRAND SYNTHESIS OF DARWIN'S THEORY

In the final recapitulation of his theory, Darwin synthesises the myriad lines of evidence that support *natural selection* as the primary mechanism driving the *evolution of species*. He presents the process as both *gradual* and *incremental*, shaped by environmental pressures that act over immense spans of time. The argument for the *mutability of species* is laid out in stark contrast to the immutability long championed by the scientific establishment, with Darwin challenging the prevailing orthodoxy and laying the groundwork for modern biology.

In this grand narrative, Darwin elevates natural selection from a mere hypothesis to a *universal principle* governing the evolution of all life, one which would forever alter humanity's understanding of its place in the natural world.

In weaving the fundamental principles of *natural selection*, *artificial selection*, and other core Darwinian concepts such as *common descent*, *sexual selection*, and *survival of the fittest* into the intricate tapestry of evolutionary theory, we delve deeper into the mechanisms that not only guide species toward adaptation but also explain the diversity and complexity of life through both *heredity* and *competition*.

ARTIFICIAL SELECTION: A HUMAN-CRAFTED MODEL FOR NATURAL SELECTION

The concept of *artificial selection*, introduced by Darwin as a model for understanding natural selection, serves as a critical lens through which we can observe how the deliberate choices of humans in breeding domesticated species lead to significant changes over relatively short periods. By selectively breeding animals such as dogs, horses, or pigeons, based on desirable traits such as size, strength, or colouration, humans have effectively accelerated evolution by *imposing selective pressures* which mirror those found in nature.

Darwin's use of *artificial selection* as an analogy is illuminating, showing that if humans can direct changes in populations through selective breeding, then nature—operating over vastly longer time scales and under constant environmental pressures—could also shape species by selecting traits which confer *adaptive advantages*. In this way, artificial selection becomes a microcosm of *natural evolutionary processes*, providing evidence for the ability of *selection forces* to drive *significant evolutionary change*.

NATURAL SELECTION: THE CORE MODEL OF EVOLUTIONARY MECHANISM

At the heart of Darwin's evolutionary theory is *natural selection*, the process by which *individuals with advantageous traits*—those traits better suited to their environment—survive at higher rates and reproduce more successfully than others. This model underscores the essential mechanism by which species evolve over time. Darwin's insight was that, in a population of organisms, there exists *variation*, and this variation means that some individuals are better equipped to survive in their environment than others. Over successive generations, these advantageous traits become more prevalent within the population.

The implications of natural selection are profound. It explains how species can adapt to their environments through a *differential survival* mechanism, where the *fitness* of an organism—defined by its ability to survive and reproduce—is determined by the interaction between its inherited traits and the surrounding environment. Natural selection is, therefore, not a force acting with intent but a *blind process* which favours traits that improve an individual's reproductive success.

THE INFANCY OV THOUGHT

COMMON DESCENT: THE TREE OF LIFE

A cornerstone of Darwin's theory is the concept of *common descent*, the hypothesis that all living species share a *common ancestor* from which they have diversified over time. This notion introduces the metaphor of the *Tree of Life*, where species diverge and branch off from common points, representing their evolutionary paths. The principle of common descent unifies the diversity of life under a single evolutionary framework, suggesting that all species, no matter how different they appear today, are connected through a long and intricate process of evolution.

Homologous structures in different species, as well as *vestigial organs*, provide compelling evidence for common descent. These structures, which may serve different functions in different species, share a common developmental origin, indicating that these species have evolved from the same ancestral form. The Tree of Life, therefore, not only represents the diversification of species but also reveals the *underlying unity* of life, showing how all organisms are part of the same evolutionary continuum.

[8]SEXUAL SELECTION: EVOLUTION BEYOND SURVIVAL

Sexual selection, a concept Darwin developed as a subset of natural selection, explains how certain traits evolve not because they provide a direct survival advantage, but because they increase an individual's *reproductive success* by making them more attractive to potential mates. Sexual selection can lead to the evolution of traits that may even be *detrimental to survival*, such as the extravagant plumage of a peacock or the large antlers of a stag. These traits are maintained in populations because individuals possessing them have *greater reproductive success*, even if the traits do not enhance their survival.

In this model, *mate choice* becomes a powerful evolutionary force, shaping species through preferences that drive the development of characteristics linked more to *reproductive desirability* than to survival. While sexual selection does not operate in isolation from natural selection, it introduces an additional layer of complexity in how traits evolve, highlighting that reproductive success, rather than mere survival, is the ultimate measure of *fitness* in Darwinian evolution.

9SURVIVAL OF THE FITTEST: DIFFERENTIAL SURVIVAL AND ADAPTATION

Survival of the fittest, a phrase famously attributed to Darwin (though originally coined by Herbert Spencer), succinctly encapsulates the competitive dynamic inherent in natural selection. In this context, "fitness" refers not to physical strength alone, but to an organism's overall ability to *survive and reproduce* in its environment. Those individuals whose traits best suit the environmental pressures they face—whether through physical adaptations, behaviours, or reproductive strategies—are the ones that are most likely to pass on their genes to the next generation.

This struggle for survival, influenced by the *Malthusian concept* of limited resources, ensures that only the fittest individuals—those that can outcompete others for food, shelter, and mates—will leave the most offspring. Over time, the population will shift, favouring those traits that enhance fitness. Thus, *natural selection* operates as a refining process, continuously sculpting populations by favouring *adaptive traits* and eliminating those that are less favourable in the given environment.

[10]INTEGRATING DARWIN'S MODELS: A COMPREHENSIVE UNDERSTANDING OF EVOLUTION

The synthesis of *artificial selection, natural selection, common descent, sexual selection,* and *survival of the fittest* provides a comprehensive understanding of how species evolve over time. Darwin's insights, though grounded in the observations of his era, have been vindicated and expanded by subsequent scientific discoveries, particularly in the fields of *genetics* and *molecular biology*. Today, we understand that the variations upon which natural selection acts are caused by *mutations* in DNA, providing a *genetic basis* for evolution that Darwin could only speculate about.

Darwin's *model of evolution* represents a paradigm shift in our understanding of life on Earth. By explaining how species adapt through a gradual process of differential survival and reproduction, Darwin provided the first cohesive framework for understanding *biological diversity*. His theory of *common descent* redefined the relationships between species, while *sexual selection* added a new dimension to the evolutionary process by explaining traits which seem, at first glance, to defy the logic of natural selection. Together, these models form a robust, multi-faceted explanation of how life evolves and adapts, grounded in the *principles of natural law* and driven by the *blind but purposeful* forces of nature.

THE INFANCY OV THOUGHT

In essence, Darwin's contributions lay the foundation for the modern biological sciences, transforming the way humanity understands its own origins and its place within the *Tree of Life*. Through the processes of *selection*, whether natural or sexual, and the branching of species from common ancestors, we see that life is in a state of continuous *evolutionary flux*—a complex, dynamic system shaped by the interplay of genetic variation, environmental pressures, and reproductive success.

The concept of *Darwinian evolution* is not merely a narrative of survival but a deeply intricate mechanism, woven with both *biological necessity* and the *pressures of existence*. It is within this context that *emotional development* and *conscientious necessitation* find their place as emergent evolutionary tools, honed through the millennia, underpinning *human social evolution*.

NEUROGENESIS

The development of *neurogenesis*, the intricate orchestration of *nervous systems*, and the *emergence of consciousness* have become focal points in the larger framework of *evolutionary biology*. Their roles, interwoven into the fabric of *Darwinian concepts*, provide a detailed and sophisticated understanding of how the *biological evolution* of species is not merely a physical transformation, but one which involves profound neurological and cognitive shifts.

The process of *neurogenesis*, or the birth of new neurons, is fundamental to the adaptive capabilities of organisms within Darwin's framework of *natural selection*. As environments shift, organisms not only undergo physical adaptations but must also develop *cognitive tools* that allow for improved *sensory perception, decision-making*, and *survival strategies*. In this regard, *neurogenesis* serves as a mechanism by which the brain can evolve and remain plastic, thus enhancing its ability to form *new neural networks* in response to changing stimuli. This process is not static but continuous, and it can be seen as one of the driving forces behind the development of *complex behaviours* in species.

In Darwin's model, the *nervous system* is pivotal in shaping the *biological outcomes* of *natural selection*. It is the *nervous system* that underpins *an organism's ability to interact with its environment* and respond to the pressures which govern *survival and reproduction*. As environmental demands shift, so too does the necessity for a *more advanced nervous system*, capable of not only rapid reflexive action but also *higher-order cognitive functions*. In species that display greater *cognitive complexity*, such as humans and higher mammals, the role of *neurogenesis* becomes paramount. The generation of new neurons and the rewiring of neural circuits allow these organisms to develop *behaviours* that go beyond mere survival—behaviours that involve *problem-solving, empathy*, and *cooperation*, traits that Darwin touched upon in his discussion of *social evolution*.

COMPLEX SYSTEMS: NERVOUS SYSTEMS

The *mammalian nervous system(s)* is an extraordinary network, a complex tapestry of interconnected components that orchestrates the coordination, perception, movement, and cognition of the organism. It is not a singular entity, but a system of systems, each with its unique functions and intrinsic applications, yet

THE INFANCY OV THOUGHT

working in seamless harmony. The nervous system is traditionally categorised into two major divisions—the *central nervous system (CNS)* and the *peripheral nervous system (PNS)*—each containing subdivisions which operate in a symbiotic relationship, ensuring the organism's survival and interaction with its environment in profoundly sophisticated ways.

The *Central Nervous System* (CNS) is composed of the *brain* and the *spinal cord*, forming the command centre of the body. It is within the CNS that higher-order functions such as *thought, reasoning, emotion*, and *memory* are processed. The CNS controls and regulates the activities of the entire nervous system, processing sensory information, coordinating voluntary and involuntary actions, and modulating the responses of the peripheral components. The *brain*, with its intricate neural networks and specialised regions, is the primary organ for *cognitive processing, consciousness*, and *decision-making*. The *spinal cord*, while often viewed as a mere conduit for signals, is a complex structure, controlling reflexes and facilitating the communication between the brain and the peripheral body.

The *Peripheral Nervous System* (PNS) is divided into two further categories: the *somatic nervous system (SNS)* and the *autonomic nervous system (ANS)*. These subdivisions allow the body to interact with both the external environment and regulate internal homeostatic mechanisms The PNS itself is a vast network of *nerves* and *ganglia* which relay information from the CNS to the muscles and glands, and from the sensory receptors back to the CNS.

The *Somatic Nervous System* (SNS) controls voluntary movements and sensory reception. It consists of *sensory neurons*, which carry information from the *sensory receptors* in the skin, muscles, and joints, as well as *motor neurons* that send signals from the CNS to skeletal muscles. These motor neurons activate *voluntary muscle contractions*, facilitating *conscious movement*, such as walking, grasping, or speaking. The SNS is the system which connects the body to its conscious awareness of the external environment, transmitting sensory input and allowing the organism to respond with deliberate actions. It is through the SNS that the organism's interaction with the world becomes tangible, and these responses are calibrated by sensory data processed within the CNS. Every motion, whether as complex as speech or as simple as raising a limb, relies on the accurate and precise interplay between sensory input and motor output within this system.

The *Autonomic Nervous System* (ANS) regulates involuntary functions, such as *heart rate*, *digestion*, *respiration*, and *pupil dilation*. The ANS is further divided into the *sympathetic nervous system* and the *parasympathetic nervous system*, both of which maintain *homeostasis* yet have opposing actions that ensure a dynamic equilibrium in the body. The *sympathetic nervous system* prepares the body for intense physical activity, often referred to as the "fight or flight" response. It increases *heart rate*, *dilates pupils*, redirects blood flow to muscles, and triggers the release of *adrenaline*, thus priming the organism for immediate action. On the other hand, the *Parasympathetic Nervous System* (PNS) restores the body to a state of calm after the stressor has passed, known as the "rest and digest" phase. It slows down the heart rate, promotes *digestion*, and encourages *energy conservation*. These two subdivisions of the ANS work in concert to ensure that the body can swiftly react to danger and, just as crucially, return to a state of rest, maintaining *physiological balance.*

Beyond these fundamental divisions, the nervous system includes *Enteric Nervous System* (ENS), or 'secondary central nervous system,' which is often referred to as the "second brain" due to its significant autonomy. Located within the walls of the *gastrointestinal tract*, the ENS governs the intricate processes of *digestion*, from the secretion of digestive enzymes to the muscular contractions that move food through the gut. The ENS operates largely independently from the CNS, although it communicates with the brain through the *vagus nerve*, influencing gut-brain interactions which are increasingly recognised as critical to emotional regulation and mental health. The *enteric system* exemplifies the deep interconnection between physical and mental health, with recent research highlighting how microbial balance and gut function affect *neurotransmitter production*, particularly serotonin, thus influencing *mood* and *cognitive function.*

The *nervous systems* also work in seamless collaboration with the *Endocrine System*, another regulatory system that relies on chemical signals—*hormones*—to control *metabolism, growth, reproduction*, and *stress responses*. The hypothalamus, a part of the *brain*, bridges the nervous and endocrine systems, directing hormonal outputs from the *pituitary gland* in response to *neural stimuli*. This connection exemplifies the nervous system's integration with other bodily systems, functioning to maintain the organism's *internal stability* and *external adaptability.*

THE INFANCY OV THOUGHT

The *neurogenesis* which occurs within the CNS, particularly in areas such as the *hippocampus*, demonstrates how the nervous system is not a static structure but a highly adaptive one. *New neurons* are generated in response to learning, memory, and environmental demands, allowing the brain to remain plastic and responsive throughout the organism's life. This process is central to *memory consolidation*, wherein experiences are encoded, stored, and later retrieved. The process of *synaptic plasticity*, wherein *synapses* strengthen or weaken over time based on activity, is at the heart of the brain's ability to learn, adapt, and evolve.

The *development of consciousness*, as an emergent property of this deeply interconnected neural network, represents the most profound and complex outcome of the mammalian nervous systems. *Consciousness* allows for *self-awareness, abstract thinking, emotional regulation*, and *social interaction*. The *prefrontal cortex*, that is heavily involved in *decision-making* and *executive function*, collaborates with deeper brain structures such as the *amygdala* (responsible for *emotional processing*) and the *hippocampus* (key to *memory formation*) to produce a unified and coherent *experience of self*. The coordination between these systems allows the organism not only to respond to the environment but to *reflect, plan,* and *navigate* complex social structures.

In an evolutionary respect, the combinatorial and juxtaposing frameworks of *neurogenesis, synaptic plasticity*, and the *development of consciousness* within the nervous systems provides a profound insight into how *mammalian evolution* has driven the sophistication of both *cognitive functions* and *emotional capabilities*.

THE INFANCY OV THOUGHT

The *nervous system* is not just an anatomical entity but a *living system* that evolves in response to the demands of the environment, constantly adapting to ensure the survival and flourishing of the organism. The combination of *autonomic regulation, voluntary control,* and *conscious reflection* creates a system that can *perceive, process,* and *respond* with unparalleled flexibility and precision.

Therefore, the seamless functioning of the *mammalian nervous system* represents a masterpiece of biological engineering. It allows for the dynamic and adaptable control of both involuntary and voluntary functions, enabling organisms to interact meaningfully with their environment, maintain *homeostasis,* and evolve their consciousness. Each subsystem, whether it is the *somatic, autonomic,* or *enteric nervous system,* plays a critical role in maintaining the overall balance and functionality of the organism, proving that the mammalian body is an organism not just of tissues and organs but of deeply interconnected *neural* and *cognitive* networks, all coalescing to produce life as we understand it.

The *nervous system's* evolution is directly tied to the *development of consciousness*, an emergent property of increasingly complex neural networks. *Consciousness*, in this context, is not a static entity but an evolving phenomenon that has grown more sophisticated as *neural structures* have become more complex. Darwinian evolution explains this progression as a natural response to the pressures of survival. Early nervous systems were likely rudimentary, enabling basic motor responses to stimuli. However, as species adapted to more complex ecological niches, the *nervous system* evolved in parallel, giving rise to *higher levels of awareness* and *self-perception*. The ability to *anticipate future events*, *plan*, and *socially interact* are all outcomes of this evolutionary trend toward *consciousness*.

It is through *natural selection* that the *nervous systems* of species have been continuously refined, ensuring that those who possess more effective *cognitive functions* were more successful in *surviving* and *reproducing*. This progression is particularly evident in the human species, where the expansion of the *cerebral cortex*, increased *synaptic plasticity*, and a more complex *neural architecture* have all contributed to the *sophistication* of *human consciousness*. This expansion is not just a byproduct of *genetic mutations* but is deeply connected to the *environmental pressures* that favoured *enhanced cognitive functions*—an evolutionary arms race where *intelligence*, *creativity*, and *emotional depth* became key factors in survival.

The integration of *emotional development* into this framework can be traced to the evolutionary advantages conferred by heightened *social awareness* and *empathy*. *Consciousness*, as it developed, enabled organisms to navigate increasingly complex social structures. As Darwin noted, the *struggle for existence* is not solely a physical battle but also a *psychological* and *emotional* one. The organisms that could *cooperate* more effectively, form *alliances*, and *empathise* with others had a distinct advantage in environments where *social cohesion* was necessary for survival. The *neurochemical processes*—governed by *neurogenesis* and *synaptic activity*—underpin these emotional responses, with *neurotransmitters* such as *serotonin* and *dopamine* playing a role in reinforcing *social bonds* and ensuring *group survival*.

The evolutionary process, thus, becomes a story of *conscious adaptation*—where the *nervous system*, through *neurogenesis* and increased *neural plasticity*, allows organisms to *think*, *feel*, and *interact* in more sophisticated ways. The *Darwinian concept* of *survival of the fittest* is therefore expanded beyond the realm of *physical adaptations* to include *cognitive and emotional evolution*. The *fittest* individuals are not merely those who are stronger or faster but those who can *adapt cognitively*, form *social connections*, and exhibit *emotional resilience*. This shift in understanding places *consciousness* and *neural evolution* at the heart of Darwin's theory, highlighting how species adapt not only in body but in mind.

Sexual selection also plays a crucial role in this neurological and emotional evolution. The *development of consciousness* allowed individuals to become more discerning in their mate choices, selecting partners not solely for physical traits but for *emotional compatibility* and *cognitive prowess*. This has profound implications for *reproductive success*, as individuals capable of forming *deeper emotional bonds* were likely to ensure the *well-being of offspring* and provide a more *stable environment* for their progeny. The *nervous system* thus becomes a tool not only for *survival* but for *reproductive success*, with *consciousness* driving *mate selection* through both *emotional* and *cognitive traits*.

NEURODEVELOPMENT OF CONSCIOUSNESS

In considering the *evolution of consciousness*, we also confront the *biological mechanisms* which underlie its emergence. The *nervous system*, through processes such as *neurogenesis*, not only adapted to environmental pressures but also began to exhibit *self-referential properties*, a key component of *conscious awareness*. The ability to *reflect*, to *self-monitor*, and to *experience emotions* became indispensable in the *social environments* that humans and other higher mammals inhabited. These *self-referential mechanisms*, in turn, provided an *evolutionary advantage*, as individuals who could anticipate *social dynamics*, understand the *intentions of others*, and *regulate their own emotions* were more likely to *thrive* in complex social hierarchies.

The *synaptic plasticity* that governs these *neurodevelopmental processes* is a direct response to the challenges presented by the environment. As species evolved, their *nervous systems* had to become increasingly *adaptable*, with the capacity to form *new connections* and *reorganise neural circuits* in response to changing conditions. This ability is most evident in species that have evolved *high levels of cognitive function*, such as *primates* and *humans*, where the *neural networks* which support *memory consolidation*, *learning*, and *problem-solving* have become critical to survival. *Consciousness*, in this sense, is the *culmination* of these adaptive processes, enabling organisms to not only react to their environment but to *anticipate* and *shape* it through deliberate action.

Ultimately, the integration of *neurogenesis, nervous systems*, and the *development of consciousness* into *Darwinian evolution* underscores the complexity of *biological adaptation*. The evolution of species cannot be understood solely in terms of *physical traits*; it must also account for the *cognitive and emotional dimensions,* that have become central to *survival* and *reproductive success. Consciousness,* far from being a mere byproduct of evolution, is an active force, shaping the way organisms interact with their environment, form social bonds, and pass on their genetic material to future generations. This intricate dance between *biology, environment,* and *cognition* is what ultimately drives the *evolutionary process,* ensuring which species continue to *adapt, thrive,* and *evolve* in the ever-changing theatre of life.

The idea of **EDER'S CONSCIENTIOUS NECESSITATION HYPOTHESIS**—where individuals feel compelled to adapt, conform, or deviate based on social requirements—is an extension of Darwin's principle of *natural selection.* In an evolutionary framework, social cohesion has been as vital to human survival as the ability to physically adapt to environmental challenges. The emotional constructs that underpin *social utility,* such as *empathy, loyalty,* and *cooperation,* have not evolved in a vacuum but through the pressures of surviving in *complex social groups.* These emotional traits have been favoured by *natural selection,* as individuals capable of *building bonds* and *cooperating* with others were more likely to succeed in *raising offspring* and *sustaining their genetic lineage.*

In the evolutionary context, *emotional development* is *bio-chemically grounded,* directly affecting reproductive success. The *neurochemical* and *hormonal interplay* which governs emotional responses—particularly those tied to *pleasure, affection,* and *protection*—has evolved to enhance *social cohesion*. This aligns with Darwin's broader vision of *natural selection*, as it is these emotions which foster group survival. *Oxytocin*, for instance, plays a critical role in *bond formation* and *trust*, while *dopamine* acts as a *reward neurotransmitter*, reinforcing *positive social interactions*.

Furthermore, *conscientious necessitation* works within *natural selection's* broader purview by shaping individuals' *behavioural responses* to their environment and social circumstances. Individuals exhibiting greater *conscientiousness*—in terms of adhering to *group norms*, *contributing* to the *social fabric*, and *protecting the collective*—were likely to be favoured in Darwinian terms, both *sexually* and *socially*. These behavioural patterns can be linked back to *emotional stability*, which was crucial in *maintaining group harmony* and ensuring *reproductive success*.

The *struggle for existence* detailed by Darwin intersects with these *psychosocial mechanisms*. As the pressures of survival intensified, individuals who could emotionally *adapt*—those who formed *strong social alliances*, exhibited *empathy*, and sought *cooperation*—were more likely to succeed than those who remained *isolated* or *emotionally detached*. This is especially critical in early human evolution, where *group dynamics* often determined *life* or *death*. The emotional faculties that allowed early hominins to navigate complex *social landscapes*—balancing between *competition* and *cooperation*—were selected for because they directly impacted *group survival*.

Sexual selection, another facet of Darwin's theory, adds another layer to this *emotional development*. *Mating behaviours*, especially those concerning *mate choice*, are directly tied to *emotional expression* and *conscientious displays*. *Alpha males* and *beta males*, for example, may have been distinguished not solely by *physical prowess*, but also by their *emotional intelligence*—their ability to *navigate social structures*, *secure alliances*, and exhibit the *behaviours* that made them *appealing* to potential mates. *Sexual selection*, then, did not merely favour the *strongest* or the *fastest*, but also those who could *demonstrate emotional strength, loyalty*, and *protectiveness*—qualities which signal the ability to *successfully raise offspring*.

The *concept of survival* and *emotional development* become intertwined when considering *speciation* and *divergence*. In populations subjected to *different environmental and social pressures*, emotions such as *fear, affection*, and *competition* took on varied roles. *Isolated populations*, cut off from their broader social counterparts, developed *unique emotional responses* and *social* structures which better suited their new circumstances. These *emotional traits*, favoured by their *environments*, slowly diverged, contributing to the *speciation process*. *Conscientious behaviour*, driven by localised survival needs, varied between these populations, leading to *distinct social constructs* and, eventually, *divergent species*.

In examining *conscientious necessitation* from a purely Darwinian perspective, it becomes evident that *emotional and moral frameworks* developed through *evolutionary necessity*. Humans are, by nature, a species that thrives on *social bonds*. The emotional systems which underpin *moral behaviour*, such as feelings of *guilt, compassion*, and *righteousness*, likely evolved as mechanisms for *social control* and *group stability*. Early humans who failed to adhere to *social norms*, or who lacked the emotional capacity to *navigate group dynamics*, would have been less successful in both *survival* and *reproductive contexts*. Over time, these emotionally-driven systems of behaviour would become *hardwired* into the human *psychosocial architecture*, a process that Darwin hinted at, though left to future generations to explore.

When integrated with *Darwinian concepts* of *variation*, *natural selection*, and *speciation*, both *emotional development* and *conscientious necessitation* serve as critical evolutionary tools that shaped the trajectory of human evolution. These faculties not only aided in *social bonding* and *cooperation*, but also played an essential role in the *selection processes* that determined which individuals thrived and which were left behind. Far from being peripheral to survival, *emotional systems* are at the core of *evolutionary success*, governing the behaviours which allowed early humans to adapt to their *environmental challenges* and *social landscapes*. Through the lens of *natural selection*, these emotional traits become crucial mechanisms for maintaining *group (and social) cohesion*, ensuring *reproductive success*, and fostering the *evolution of complex societies* that continue to shape human existence today.

THE INFANCY OV THOUGHT

CH. 3: A.D. AFTER DARWIN

EDERIAN HYPOTHESES, BUILT UPON THE FOUNDATIONAL PRINCIPLES OF DARWINIAN BIOLOGY

- CONTEXTUAL CORRECTION FOR HUBBLE'S CONSTANT
- EDER'S EXPONENTIAL EXPANSION MODEL
- CONTEXTUAL CORRECTION FOR HUBBLE'S CONSTANT EDER'S EXPONENTIAL EXPANSION MODEL
- ABORIGINAL HYPOTHESIS
- THE ABORIGINAL PROGENITOR HYPOTHESIS
- 'WILLED MOTION' ARGUMENT
- EDER'S "ATOM TO ADAM," (FROM ATOM, CAME ADAM) HYPOTHESIS
- EDER'S SEXUAL AFFIRMATION HYPOTHESIS
- EDER'S REFINED REFLEXIVITY AND NEURONAL EXCITATION HYPOTHESIS
- EDER'S ATMOSPHERIC CONDITIONING HYPOTHESIS
- BIOLOGICAL REPRODUCTION, MUTATION, AND RESTRUCTURING
- THE EVOLUTIONARY ORIGINATION OF HOMOSEXUALITY HYPOTHESIS

THE INFANCY OV THOUGHT

- THE HONEYMOON PHASE, IN CYCLICALITY HYPOTHESIS

THE INFANCY OV THOUGHT

EDERIAN HYPOTHESES

Abstract

The current understanding of the universe's expansion is governed by Hubble's Law. Wherever, recent observations suggest an exponentially increasing rate of expansion. This paper introduces a novel equation, 'Eder's Exponential Expansion Model', which accounts for this phenomenon by incorporating a time-dependent Hubble parameter. The proposed model provides a more accurate representation of the universe's growth over time.

Introduction:

The discovery of the universe's expansion by Edwin Hubble revolutionised our understanding of cosmology. Hubble's Constant has long been used to describe this expansion. However, new evidence suggests the rate of expansion is not constant but increasing exponentially. This paper presents a new mathematical model, 'Eder's Exponential Expansion Model', which better reflects these observations. The derivation and implications of this model are discussed in detail.

In this study, I propose a novel equation to represent the exponential growth of the universe's scale factor, considering an exponentially increasing Hubble parameter. This equation is hereby referred to as 'Eder's Exponential Expansion Model'."

THE INFANCY OV THOUGHT

Equation Presentation:

$$a(t) = a_0 \exp\left(H_0/k \; ^{[\frac{H0}{K}]} (e^{kt} - 1)\right)$$

The cosmological expansion described here involves an exponential growth model, influenced by Hubble's Law. Hubble's Law states that the rate at which a galaxy moves away from us (its recessional velocity, v) is proportional to its distance from us (d), expressed as $v = H_0 d$ where H_0 is the Hubble constant.

To represent the expansion of the universe mathematically, we consider the scale factor $a(t)$, which describes how distances in the universe change with time. For an exponential expansion, the scale factor $a(t)$ grows exponentially over time.

The general form of the scale factor in an exponentially expanding universe is:

$$a(t) = a_0 e^{H0t}$$

where:

- a_0 is the scale factor at the initial time (usually set to 1 at the present time),
- h_0 is the Hubble constant,
- t is the time since the initial moment (Big Bang).

Given the description, where the rate of expansion increases exponentially every millionth of a second, we can adapt the model accordingly. Suppose h_0

represents the base Hubble constant (in units per second). The rate of expansion would then scale as:

$H(t) = H_0 \cdot e^{kt}$

where:

- $H(t)$ is the time-dependent Hubble parameter,
- k is a constant representing the rate of exponential increase in the Hubble parameter,
- t is time in seconds.

To find the proper model over the entire existence of the universe (estimated at about 13.8 billion years, or roughly 4.354×10^{17} seconds), we integrate this model over time.

First, we recognise that a(t)a(t)a(t) depends on the integral of $H(t)$: $a(t) = a_0 e^{\int_0^t H(t') \, dt'}$

Given $H(t) = H_0 e^{kt}$, we have: $\int_0^t H(t') \, dt' = \int_0^t H_0 e^{kt'} \, dt' = H_0/k \, [\frac{H_0}{K}] \, (e^{kt} - 1)$

Thus, the scale factor becomes: $a(t) = a_0 \exp(H_0/k \, [\frac{H_0}{K}] \, (e^{kt} - 1))$

This equation captures the exponential growth of the universe's scale factor considering an exponentially increasing Hubble parameter over time.

In summary, the mathematical model representing the universe's expansion, with an exponentially increasing rate of expansion every millionth of a second, is:

THE INFANCY OV THOUGHT

$$a(t) = a_0 \exp\left(H_0/k^{\left[\frac{H0}{K}\right]} (e^{kt} - 1)\right)$$

where H_0 is the base Hubble constant, k is a constant representing the exponential growth rate, and t is the time in seconds.

Given *that* entropy is related to the beginning point of the Big Bang, we need to consider the context in which entropy (S) and Hubble's constant (H_0) are used. The entropy of the early universe can be extremely high due to the large number of particles and energy states involved.

Given Expression

$$10^6 \cdot 10^6 \cdot \frac{1}{10^{-10}} H_0$$

$$S = 10^{22} \cdot H_0 \mid S \leq 10^{22} \cdot H_0 \mid$$

...which leads to...

The *Aboriginal Hypothesis*, as conceived within the context of human evolution and natural selection, posits a profound and complex intersection between biological evolution, cultural development, and the subconscious undercurrents that guide human behaviour and thought. It draws on the recognition that *natural selection*, long seen as the driving force behind the adaptation of species to their environments, does not operate solely at the genetic or phenotypic level but extends into the realms of human consciousness, culture, and societal evolution. As we consider the evolution of *Homo sapiens*, the process that brought humanity from *Homo habilis* through *Homo erectus* to modern man, we must question whether the same mechanisms of *variation*, *selection*, and *adaptation* which shape biological organisms also apply to the collective and unconscious choices we make as a species.

NATURAL SELECTION AND THE HUMAN EVOLUTIONARY PATHWAY

The sequence of human evolution, from *Homo habilis*—the earliest form of man who lived between 1.5 to 2 million years ago, possessing primitive tool-making capabilities—to *Homo erectus*, who emerged approximately 500,000 years ago and marked the advent of true bipedalism, reveals an ongoing process of adaptation. This evolutionary journey continued with the rise of *Neanderthals* around 100,000 years ago, who demonstrated rudimentary culture, group living, and tool use, before the emergence of *Homo sapiens* around 40,000 years ago. The shift from *Homo erectus* to modern man is marked not only by physical adaptations but by significant cognitive developments that allowed for advanced tools, language, and social structures.

However, the hypothesis suggests that natural selection in humans has become intertwined with an *unconscious self-selection* process. Unlike other species, where evolution is driven purely by environmental pressures, humans—through culture, intellect, and societal structures—have become active agents in their own evolution. Through mechanisms such as *cultural selection*, *social norms*, and *psychosocial dynamics*, humans may be selectively breeding themselves, albeit unconsciously, toward traits that favour *social agreeableness*, *compliance*, and *utility* within a given cultural or societal framework.

SELF-SELECTION AND THE COLLECTIVE UNCONSCIOUS

The notion that *humans* may unconsciously breed or select themselves for specific traits implies that, beyond the natural selection Darwin described, there is an internal, self-guiding process at play. This hypothesis proposes that our *collective unconscious*—a term borrowed from Jungian psychology—acts as a hidden driver of evolution. This *unconscious breeding* selects for traits that may serve to enhance group cohesion or social functionality but, paradoxically, may also lead to stagnation or devolution in other aspects of intellectual or emotional development.

For instance, as humans continue to group themselves based on *cultural*, *ideological*, and *philosophical similarities*, we engage in a form of self-selection for *agreeableness* and *compliance* within those groups. Evolution, in this sense, becomes a *feedback loop*, where social behaviours and cultural traits influence reproductive patterns, leading to a population that increasingly reflects the values and biases of its environment. This subconscious breeding process may result in a *homogenisation* of traits within populations, favouring conformity over individualism and innovation.

Yet, this self-selection does not necessarily lead to the most optimal evolutionary outcomes. As the hypothesis posits, humans may be breeding themselves into *intellectual complacency*, resulting in the perpetuation of *simplistic thought patterns* and *cultural biases*. The *crowd mentality* and *societal norms* that shape human interactions may inadvertently stifle the very traits that drive *intellectual evolution*, such as *critical thinking*, *creativity*, and *individualistic thought*. This paradox is evident in the modern world, where the pressure to conform to cultural or societal standards often overrides the pursuit of *intellectual independence* and the *objective evaluation of ideas*.

EVOLUTIONARY STAGNATION AND THE PARADOX OF INTELLECTUAL IMMATURITY

The Aboriginal Hypothesis suggests that *intellectual stagnation* and the *paradox of human development* arise from this very dynamic of self-selection. Despite the potential for individualism and the capacity for *critical reasoning*, humans as a species seem to exhibit a tendency toward *intellectual immaturity*, remaining trapped in cyclical patterns of *cultural bias*, *groupthink*, and *emotional volatility*. These traits, once essential for survival in early human history, now appear as evolutionary relics that hinder further progress. The *hormonal drives*, such as *lust* and *aggression*, that once facilitated survival and reproduction in ancestral environments now act as obstacles to intellectual and social advancement.

The hypothesis argues that the evolutionary forces that once selected for traits such as *selfishness* and *aggressiveness* are no longer beneficial in the current era. These traits, once necessary for survival in a harsh and competitive world, now serve to perpetuate conflict, division, and intellectual regression. Humanity's failure to evolve beyond these primitive impulses has led to a *psychophysiological imbalance*—an inability to transcend base instincts in favour of *higher reasoning* and *altruism*.

THE NECESSITY OF FORCED EVOLUTION: TRANSCENDING NATURAL SELECTION

To escape this evolutionary *cul-de-sac*, humanity must look beyond the traditional mechanisms of *natural selection* and consider the possibility of *forced evolution*—an intentional manipulation of our own genetic, cognitive, and social systems to overcome the inherent limitations of natural evolution. This could take the form of *genetic engineering*, *biotechnological enhancement*, or the *harnessing of artificial intelligence* to augment human cognitive capacities. Such interventions, while ethically complex, may be necessary to propel humanity into the next phase of *intellectual and emotional development*.

Forced evolution would allow for the deliberate selection of traits which favour *intellectual maturity*, *emotional stability*, and *social cohesion* without the need for unconscious or naturalistic processes. This would involve a conscious effort to break free from the limitations imposed by our evolutionary past and to embrace a future where humanity is no longer subject to the random and often inefficient forces of *natural selection*. By selectively breeding or engineering for traits which promote *altruism, critical thinking*, and *intellectual independence*, humanity could potentially overcome the paradox of intellectual immaturity and achieve a new level of *evolutionary sophistication*.

THE ETHICAL DILEMMA: MANIPULATING HUMAN EVOLUTION

However, the notion of *forced evolution* raises significant ethical concerns. Is it morally acceptable to intervene in the natural evolutionary process, if only to accelerate human development? Does the pursuit of *intellectual evolution* justify the potential risks associated with genetic or technological manipulation? These questions are not easily answered, but they must be considered as humanity faces the prospect of stagnation.

In the absence of natural selection providing the necessary pressures for continued evolution, it may be argued that forced evolution is not only ethical but *necessary* for the survival and advancement of the species. Without intervention, humanity risks falling into a state of *devolution*, where the traits that once served us become the very traits that hold us back. *Selfishness*, *aggressiveness*, and *intellectual complacency* could become the hallmarks of a species that is no longer capable of adapting to the complexities of the modern world.

In essence, the *Aboriginal Hypothesis* proposes the only way forward is through the *conscious manipulation* of our own evolution. By taking control of the evolutionary process, humanity can break free from the constraints of its biological and intellectual past and move towards a future defined by *collective consciousness*, *intellectual maturity*, and *ethical altruism*.

THE INFANCY OV THOUGHT

The Aboriginal Progenitor Hypothesis

The events leading to the formulation of the *Aboriginal Progenitor hypothesis*, and the subsequent exploration of Darwinian fallacies within the framework of human cognition and evolution, must be elucidated through an in-depth analysis of the intersections between *biology*, *philosophy*, and the *epistemological errors* present in the foundational frameworks of evolutionary theory. In this inquiry, we begin by addressing the inherent limitations within Darwin's method, specifically his subjective omissions and the resulting gaps that, in retrospect, can be seen as *epistemological fractures* in the continuity of his theoretical framework.

THE EPISTEMOLOGICAL FAILURES OF SELECTIVE OMISSION

In the canon of Darwin's work, there exist moments where critical aspects of evolutionary biology are either omitted or insufficiently elaborated upon, not because of empirical evidence that contradicts them, but because of Darwin's own subjective interpretations—his feelings of what was important or not. This selective omission results in *epistemological biases* which, left unchecked, become integrated into the collective understanding of natural selection. Such omissions, in retrospect, expose the limitations of Darwin's interpretive methodology, where the subjective viewpoint of the observer—the scientist—introduces

THE INFANCY OV THOUGHT

unavoidable biases that can affect the objectivity of the scientific process.

The point to emphasise here is not merely the subjective oversight of details but the *perils of conflating personal judgement with empirical validation*. Darwin, in omitting these details, inadvertently positioned himself as an unassailable authority within the evolutionary discourse. This act of omission, deliberate or not, raises an important question: What truths remain undiscovered, unexamined, because the scientific framework which was presented as complete is, in fact, a product of incomplete observation? Such selective omission is reminiscent of a herdsman counting sheep but choosing, intentionally or not, to skip certain members of the flock—an action that distorts the reality of what is present.

THE NECESSITY OF EVOLUTIONARY INVARIANCE

In the realm of evolutionary theory, where *natural selection* operates as the bedrock principle, the requirement for *invariance* is paramount. A theory as comprehensive and universal as evolution must be able to withstand the rigours of both empirical validation and *logical coherence*. Any deviation from this principle—any inconsistency, any subjectivity—undermines the reliability of the theory as a whole. To that end, *invariance* must extend not only to the biological processes themselves but also to the intellectual rigor applied in interpreting these processes.

Darwin's theory, while revolutionary, contains within it these epistemological weak points. It is essential that *scientific theories* like evolution remain *free from personal bias*, remaining instead grounded in empirical facts and reproducible observations. Any tendency towards omission or subjectivity risks introducing *logical fallacies* which can reverberate through the collective understanding of evolutionary biology, thereby distorting the truth. Evolution, as a *universal law of nature*, must be held to the same standards of consistency and invariance as any fundamental physical law. *Variation* within species is natural; *inconsistency* in the scientific method is not.

THE EVOLUTION OF PHYSIOLOGY: A SYMBIOTIC RELATIONSHIP WITH COGNITIVE FUNCTION

The second half of this analysis deals with the *reciprocal evolution* of the human body and brain, and how the body must evolve to sustain the heightened *cognitive demands* of an evolving *psychophysiology*. As human beings develop greater intellectual capacity, increasing in both *cognitive complexity* and *consciousness*, the brain expands in both size and capability. Such a transformation is not without its physical consequences—the physiological structure of the human body must change to accommodate the needs of this increasingly sophisticated brain.

Consider the example of *extra-sensory perception (ESP)* as a speculative extension of cognitive evolution. In the hypothetical scenario where the human brain develops enhanced perceptive abilities beyond the five classical senses, the body would likely undergo parallel physical changes to support these new cognitive functions. For instance, if humans evolved the ability to perceive and interact with higher-dimensional space, or if *telepathic communication* became possible through newly developed neural pathways, the body would no longer need certain physical features associated with our current modes of perception. The evolution of these new abilities might render certain physiological traits—such as *toes for balance*—obsolete.

Similarly, as the brain's demand for *protein* and *nutrients* increases, the physiological structure must adapt to provide these resources more efficiently. The larger the brain becomes, the more dependent the organism becomes on a *protein-rich diet*, driving both the *physiological and behavioural evolution* of the species. In this sense, the body is in constant *adaptive response* to the evolving demands of the brain, and the two cannot be understood in isolation. The *cognitive evolution* of humanity is intrinsically linked to our *physiological evolution*, a dynamic interplay that shapes the future of our species.

THE ABORIGINAL PROGENITOR AND THE QUEST FOR CONSISTENCY

The *Aboriginal Progenitor hypothesis* asserts that the continuity of evolutionary theory, particularly as it relates to the evolution of *consciousness*, requires a rigorous adherence to the principle of *consistency*. In Darwin's framework, we see a tendency toward *inconsistency*, particularly in the omission of critical variables that may challenge the initial premise of natural selection. This inconsistency introduces *fallibility* into a theory that must be held to a higher standard of *invariance* if it is to continue as a foundational pillar of biological understanding.

For a hypothesis such as *evolution* to hold absolute scientific weight, it must be *unerring* in its application across all levels of life—genetic, physiological, and cognitive. The Progenitor theory posits that human evolution, particularly the evolution of our *psychophysiology*, is not merely a matter of biological change but an ongoing process that reflects deeper *cosmic principles* of order, complexity, and *self-awareness*. As we evolve, the question becomes: Can we develop a framework of *evolutionary invariance* that accounts for not only the biological but also the *cognitive* and *metaphysical* dimensions of our existence?

THE COSMIC EVOLUTIONARY ARCHETYPE: BEYOND DARWIN

The Aboriginal Progenitor hypothesis transcends the limitations of Darwin's initial framework by considering the *cosmic implications* of evolutionary theory. Evolution, in this expanded view, is not merely a biological process confined to the physical world; it is a *cosmic principle* that governs the emergence of *consciousness* and *self-awareness* throughout the universe. The notion of a *cosmic progenitor*—a *bornless intelligence* that initiates the unfolding of the universe—reframes the process of evolution as a *self-organising principle* that applies not only to biological organisms but to the *fabric of reality* itself.

In this context, the evolution of human consciousness is a microcosmic reflection of the *greater evolutionary process* that began with the *Big Bang*. The expansion of the universe, the emergence of *order from chaos*, and the continual process of *self-organisation* reflect the same principles which drive biological evolution. Human beings, in their evolution from *Homo habilis* to *Homo sapiens*, are participating in a *larger cosmic process*—one which transcends the limitations of *time* and *space*.

This expanded framework challenges the limitations of Darwin's work by incorporating a *multidimensional view* of evolution—one that is not constrained by the physical processes of natural selection, but which accounts for the *emergence of consciousness* as a fundamental aspect of cosmic evolution. The *Aboriginal Progenitor* becomes not only the initiator of biological evolution but the *cosmic architect* of consciousness, guiding the evolutionary process through a series of *self-perpetuating principles* that exist beyond the limitations of Darwin's original theory.

TOWARDS A UNIFIED THEORY OF EVOLUTION AND CONSCIOUSNESS

The exploration of Darwin's epistemological limitations and the Aboriginal Progenitor hypothesis calls for a *revolutionary rethinking* of evolutionary theory. *Evolution* must no longer be seen as a purely biological process but as a *multidimensional phenomenon* that incorporates the evolution of *consciousness, intellectual capacity*, and *cosmic self-awareness*. The failure to acknowledge the *invariance* of these processes—whether through subjective omission or epistemological bias—undermines the *universality* of evolution as a cosmic principle.

As humanity continues to evolve, the need for a new, unified theory—one which bridges the gap between biological evolution and the *cosmic evolution of consciousness*—becomes increasingly apparent. The Aboriginal Progenitor hypothesis offers a pathway toward this *integration*, suggesting that our continued evolution is not merely biological but also deeply connected to the *self-organising principles* which govern the universe. This new framework, free from the limitations of *inconsistency* and *subjective bias*, will provide a more profound understanding of the forces that shape both *humanity* and the *cosmos* itself.

Coevolution is one of the most intricate and powerful processes driving the diversification and adaptation of life on Earth. This dynamic interaction between species, where evolutionary changes in one species drive reciprocal changes in another, is essential for maintaining *biodiversity* and ensuring the resilience of ecosystems. The constant evolutionary dialogue between predators and prey, parasites and hosts, plants, and pollinators, and even between competing species, shapes the ecological fabric, ensuring that species are continually adapting to shifting environments. This concept, though widely discussed in evolutionary biology, is explored with novel insights by *Jacob A. Eder*, who emphasises the critical role of *coevolutionary pressures* in fostering *beneficent mutations*—within his **'WILLED MOTION' ARGUMENT** those that enable organisms to adapt to *environmental shifts* and *geological changes* in ways that are ultimately advantageous to the survival of species.

At its core, *coevolution* acts as a *reciprocal evolutionary mechanism*, where the evolutionary change in one species acts as a selective force on another, which, in turn, responds with its own evolutionary adaptations. This interaction often manifests in highly specialised relationships, where the *fitness of both species* becomes interdependent. One of the most striking examples of coevolution is the relationship between *flowering plants and their pollinators*, where plants evolve specific traits such as nectar, colour, scent, and shape to attract pollinators, while the pollinators simultaneously evolve traits which allow them to efficiently access the plant's resources. Over time, these *coevolutionary arms races* generate extraordinary *biodiversity*, as each adaptation by one species drives further specialisation in the other.

The *necessitation of biodiversity* through coevolution lies in the fact that each organism, by adapting to its environment and other organisms within it, creates *new ecological niches* that other species can exploit. This constant branching of ecological opportunities encourages *speciation* and further *ecological diversification*, resulting in a more complex and interconnected web of life. Coevolution thus acts as a powerful catalyst for biodiversity, as species continually evolve to better exploit their resources or defend themselves from other organisms. In this way, *biotic interactions*, such as competition, predation, and symbiosis, become *selective pressures* that sculpt the evolution of species.

Environmental shifts—whether through *geological changes*, such as tectonic movements, volcanic activity, or climate shifts—create *new ecological challenges* that organisms must adapt to if they are to survive. These changes alter the landscape, the availability of resources, and the nature of interactions between species, all that drive *coevolutionary responses*. As environmental conditions change, mutations which may have been neutral or even slightly deleterious in a stable environment may suddenly become advantageous, allowing the organisms carrying those mutations to exploit new niches or better compete within their existing ones.

Beneficent mutations, in this context, refer to those genetic changes that confer a survival advantage, particularly in response to *coevolutionary pressures* and *environmental shifts*. The coevolutionary process increases the likelihood that such mutations will arise, as the constant *adaptive pressure* from interacting species keeps populations in a state of *genetic flux*. Organisms which can quickly adapt to the changes in their environment, often facilitated by these mutations, are more likely to survive and reproduce, passing on these advantageous traits to subsequent generations.

The relationship between *coevolution* and *biodiversity* becomes especially clear when considering how *geological changes* reshape ecosystems. The *fragmentation* of habitats due to tectonic activity, the formation of mountains or islands, or the drying up of lakes and rivers can isolate populations, leading to *allopatric speciation*. These geological shifts not only separate populations but also create *novel environments* that impose new selective pressures on the organisms that inhabit them. Isolated populations must adapt to the unique conditions of their new environments, and in doing so, they accumulate genetic changes, including beneficial mutations that improve their chances of survival.

In addition to direct environmental pressures, these isolated populations may also face new *biotic interactions*, that drive further coevolutionary dynamics. A population separated from its ancestral home may encounter new predators, competitors, or symbiotic partners. The selective pressures exerted by these new interactions encourage the evolution of *novel adaptations*, contributing to the biodiversity of the ecosystem. Over time, the evolutionary divergence between isolated populations results in the emergence of new species, each finely tuned to its ecological niche.

A particularly fascinating example of how *coevolution* and *environmental changes* drive *beneficent mutations* can be seen in *pollination syndromes*. In regions where climate shifts alter the availability of pollinators, plants may evolve new traits to attract different pollinator species or even shift to wind pollination if animal pollinators become scarce. In turn, pollinators may evolve traits that allow them to exploit a broader range of floral resources. These changes, driven by both *abiotic factors* (such as climate) and *biotic interactions* (such as mutualism), increase the genetic variability of both plants and pollinators, providing the raw material for natural selection to act upon.

This exploration introduces the idea that *coevolutionary pressures* not only lead to advantageous mutations but also serve as a buffer against extinction during periods of *environmental upheaval*. Species that have evolved through *coevolution* are often better equipped to adapt to sudden environmental changes because their evolutionary history has been shaped by *constant adaptation*. For instance, organisms that have coevolved in *predator-prey dynamics* tend to evolve *flexible defence mechanisms*—such as the ability to alter their behaviour, physiology, or even physical traits in response to predator pressures. These traits, initially evolved in response to other species, may inadvertently confer advantages in the face of broader environmental changes, ensuring that *coevolved species* maintain *adaptive potential* in unstable environments.

Moreover, *coevolutionary dynamics* tend to promote *genetic diversity* within populations, as different individuals may evolve different strategies to cope with the selective pressures imposed by other species. This genetic diversity serves as a reservoir for future *adaptive potential*, allowing species to weather environmental shifts more effectively than populations with limited genetic variation. By fostering this *genetic variability*, coevolution promotes the *long-term survival* of species, ensuring that life can continue to thrive even in the face of geological upheavals and climate change.

The *Red Queen Hypothesis*, which describes the constant evolutionary "arms race" between interacting species, encapsulates the essence of coevolution. The hypothesis suggests that species must continuously evolve not just to gain an advantage but to *maintain their current fitness* relative to the species with which they interact. This constant coevolutionary pressure keeps populations in a state of *evolutionary readiness*, ensuring that when environmental shifts occur—whether geological or climatic—species are more likely to possess the *genetic diversity* and *adaptive traits* necessary for survival. Beneficent mutations, in this sense, are not isolated anomalies but the *inevitable outcome* of millions of years of reciprocal evolutionary pressure, always pushing organisms towards greater *adaptive complexity*.

The *coevolutionary web*—the interconnectedness of species and their environments—is not merely a backdrop to evolution but a *driving force* behind the diversity of life. The *necessitation of biodiversity* through coevolution is not simply a byproduct of random mutations and selection pressures but a *deliberate* consequence of how life evolves in tandem with itself. In this way, coevolution does not just shape individual species but the entire *ecosystem*, creating a *self-sustaining system* where the continued evolution of one species depends on the evolution of another, ensuring that biodiversity remains not only a byproduct of life's complexity but its essential foundation.

The coevolutionary process is indispensable to the diversification of life. It acts as the *causal mechanism* behind much of the biodiversity we observe today, linking *genetic mutations, environmental shifts*, and *species interactions* in a continuous loop of reciprocal adaptation. Coevolution is the force which generates *beneficent mutations* in response to both *biotic pressures* and *abiotic changes*, ensuring that life remains dynamic, adaptable, and diverse in the face of environmental flux. It is through coevolution that species not only survive but thrive, contributing to the complex and interconnected web of life that balances the ecological structure of our planet.

Willed Motion, the conscious direction of movement as executed by an organism, represents one of the most complex phenomena in biological and neurological systems. It is the culmination of countless integrative processes, spanning multiple domains of cellular biology, electrophysiology, neurochemistry, and biomechanics. To understand willed motion is to engage with a multilayered symphony of neural, muscular, and sensory feedback loops, all under the guidance of higher cortical functions. The emergence of deliberate, goal-directed movement from the abstract impulses of the mind is not merely an electrochemical cascade but a testament to the evolutionary sophistication of the human nervous system, particularly in its hierarchical coordination of motor commands, decision-making, and sensory integration.

THE CORTICAL INITIATION OF WILLED MOTION: FROM THOUGHT TO ACTION

At the heart of willed motion lies the cerebral cortex, specifically the primary motor cortex (M1), located in the precentral gyrus of the frontal lobe. Here, the generation of volitional movement begins with the activation of pyramidal neurons, that project directly to lower motor neurons via the corticospinal tract. However, this straightforward pathway belies the complexity underlying the initiation of movement, as willed motion is not a mere reflexive act, but one driven by conscious intent and informed by a host of sensory inputs and higher-order cognitive processes.

The prefrontal cortex, seated anterior to M1, plays a pivotal role in the conceptualisation of motion. It is within this region that decisions are made, goals are set, and strategies for movement are formulated. The prefrontal cortex integrates sensory information from the environment and internal states, evaluating potential actions through executive functions such as planning, foresight, and judgement. Once a course of action is decided, the prefrontal cortex communicates with the premotor cortex and supplementary motor areas (SMA), regions critical for sequencing complex movements and preparing the body for the execution of willed motion.

It is important to emphasise that this initial cortical phase represents the intentionality of motion—the abstract, cognitive desire to move. However, this intent must be translated into specific motor commands, a process which requires the involvement of basal ganglia circuits and the cerebellum, two regions that modulate and fine-tune motor output, ensuring that movements are smooth, coordinated, and appropriate for the task at hand.

THE BASAL GANGLIA: GATEKEEPERS OF VOLITION

The basal ganglia serve as a crucial regulatory network that modulates motor activity, acting as a gatekeeper for the flow of motor commands from the cortex to the spinal cord. These deep, subcortical structures—comprising the striatum, globus pallidus, subthalamic nucleus, and substantia nigra—are integral to the decision-making processes underlying willed motion. They function by either facilitating or inhibiting motor signals depending on the context and goal of the action. Through a process known as disinhibition, the basal ganglia selectively permit the execution of desired movements while suppressing competing motor programs, thus ensuring the precision of volitional motion.

This delicate balancing act within the basal ganglia is governed by the interplay of excitatory and inhibitory signals mediated by neurotransmitters such as dopamine, GABA, and glutamate. Dopaminergic neurons, particularly those in the substantia nigra pars compacta, are critical in modulating the output of the basal ganglia, reinforcing specific motor plans while dampening unnecessary or conflicting ones. It is within this circuitry that disorders of movement, such as Parkinson's disease, manifest, highlighting the essential role of the basal ganglia in the fluid execution of willed motion.

CEREBELLAR CONTRIBUTION: PRECISION AND COORDINATION

While the basal ganglia play a role in initiating and selecting motor programs, the cerebellum is responsible for ensuring the precision, timing, and coordination of those movements. The cerebellum receives a continuous stream of sensory information from the proprioceptive system, visual inputs, and vestibular feedback, allowing it to make real-time adjustments to motor output. Through its connections with the motor cortex and brainstem nuclei, the cerebellum compares the intended movement (the efferent copy) with the actual performance of the movement, correcting for any deviations through a process known as error correction.

The cerebellum's contribution to willed motion is most evident in the fine-tuning of complex, multi-jointed movements that require precise timing and synchronisation. Whether it be the delicate act of playing a musical instrument or the precise coordination required in athletic performance, the cerebellum is instrumental in refining and optimising the motor commands generated by cortical structures. This refinement is achieved through the modulation of Purkinje cells, which integrate vast amounts of sensory and motor information to deliver finely tuned inhibitory signals to motor pathways.

THE ROLE OF SPINAL CIRCUITS: CONDUITS OF MOTOR COMMANDS

THE INFANCY OV THOUGHT

Once the motor commands are finalised by the cortex, basal ganglia, and cerebellum, they must be transmitted to the peripheral muscles via descending spinal tracts. The corticospinal tract, which originates in the motor cortex, serves as the primary conduit for these signals, terminating on alpha motor neurons within the spinal cord. These motor neurons, often described as the final common pathway, directly innervate skeletal muscle fibres, converting the electrical signals from the brain into mechanical contractions which produce motion.

However, the spinal cord is not merely a passive relay for cortical commands; it is home to intricate networks of interneurons that contribute to the execution and modulation of movement. These spinal circuits can generate reflex arcs and rhythmic motor patterns independent of cortical input, a phenomenon known as central pattern generation. This allows for the execution of automatic, rhythmic movements—such as walking or running—without the need for continuous conscious oversight. In this sense, the spinal cord represents a lower tier of the motor hierarchy, capable of executing complex motor patterns even in the absence of direct cortical control.

NEUROTRANSMISSION AND SYNAPTIC PLASTICITY: THE MOLECULAR UNDERPINNINGS OF WILLED MOTION

At the molecular level, the execution of willed motion is dependent upon the precise transmission of electrical signals across neural synapses. This process begins with the propagation of an action potential along the axon of a motor neuron, culminating in the release of neurotransmitters into the synaptic cleft. These neurotransmitters—most notably acetylcholine at the neuromuscular junction—bind to specific receptors on the postsynaptic membrane, initiating a cascade of events that lead to the depolarisation of the muscle membrane and the subsequent contraction of muscle fibres.

The regulation of synaptic activity is not a static process but is subject to synaptic plasticity, the ability of synapses to strengthen or weaken in response to patterns of activity. Long-term potentiation (LTP) and long-term depression (LTD) are two such mechanisms that modulate synaptic strength, with profound implications for the learning and refinement of motor skills. Through repeated practice and motor learning, the neural circuits involved in willed motion undergo structural and functional changes, enhancing the efficiency and accuracy of movement over time. This phenomenon, akin to the development of muscle memory, reflects the dynamic adaptability of the nervous system and its capacity to refine motor commands through experience.

PROPRIOCEPTIVE FEEDBACK: CLOSING THE LOOP ON WILLED MOTION

The execution of willed motion would be incomplete without the feedback mechanisms that allow the brain to monitor and adjust motor performance in real-time. Proprioception, the body's sense of position and movement, is crucial in this regard, providing continuous updates to the central nervous system regarding the position, velocity, and force of the limbs. This sensory information is conveyed via muscle spindles, Golgi tendon organs, and joint receptors, which relay signals through afferent neurons to the spinal cord and brainstem, where they are integrated with motor commands.

Proprioceptive feedback allows for the rapid correction of motor errors, ensuring that movements are fluid, coordinated, and responsive to changes in the environment. For example, when walking across uneven terrain, proprioceptive inputs inform the brain of changes in limb position, allowing for immediate adjustments in gait and posture. This feedback loop is critical for maintaining balance, coordination, and precision in willed motion, highlighting the interplay between sensory and motor systems in the execution of volitional movement.

THE EVOLUTIONARY SIGNIFICANCE OF WILLED MOTION

The capacity for willed motion has profound implications for the survival and adaptation of species. In the context of human evolution, the development of fine motor control and bimanual dexterity has enabled the use of tools, the construction of complex societal structures, and the ability to manipulate the environment in ways that have afforded Homo sapiens an unparalleled evolutionary advantage. The refinement of neural circuits responsible for goal-directed movement has facilitated the development of sophisticated cognitive processes, as the planning and execution of complex motor actions are intimately tied to higher-order thinking and problem-solving.

From an evolutionary perspective, the ability to consciously direct movement represents a pivotal advancement in the adaptation of organisms to their environments. The evolution of intricate neural networks and motor pathways has allowed for the development of behaviours that are not purely reflexive but are guided by conscious decision-making and strategic planning. This has enabled organisms to engage in complex social interactions, evade predators, and secure resources, all of which are essential for survival in a dynamic and competitive ecological landscape.

The orchestration of willed motion is an awe-inspiring testament to the complexity of the nervous system. From the abstract intentions formulated in the prefrontal cortex to the fine-tuned execution of movement via spinal circuits and motor neurons, the journey of volitional movement is one of remarkable precision and coordination. It is through the seamless integration of sensory feedback, motor planning, and neural modulation that organisms can navigate their environments with agility and purpose. As we continue to unravel the intricacies of the brain and its control over the body, the profound elegance of willed motion stands as a cornerstone of our understanding of both the mind and the body.

THE INFANCY OV THOUGHT

EDER'S "ATOM TO ADAM," (FROM ATOM, CAME ADAM) HYPOTHESIS:

The phrase *"From Atom, Came Adam"* elegantly bridges the realms of science and spirituality, intertwining the fundamental building block of all matter—the atom—with the biblical narrative of man's creation. In this phrase, *the atom*, with its 6 *protons*, 6 *neutrons*, and 6 *electrons*, becomes a symbol of the divine blueprint for humanity. The atomic structure, central to all physical existence, carries with it the profound implication that *man* himself is constructed from the very essence of the universe.

In the *Book of Revelation*, the number 666 is famously referred to as the *"number of a man,"* a cryptic symbol which has elicited a myriad of interpretations across religious, historical, and esoteric texts. While traditionally seen in a more ominous context, the connection of this number to *man's creation* through the lens of science offers a different perspective. The atom, comprised of *6 protons, 6 neutrons, and 6 electrons*, represents the very building block of life as we understand it. This *numerical parallel* between the atomic structure and the number attributed to man in biblical prophecy is not mere coincidence but can be seen as a deeper *symbolic link* between *physical creation* and *spiritual essence*.

THE ATOM: FOUNDATION OF THE UNIVERSE AND MAN

THE INFANCY OV THOUGHT

The atom is the smallest unit of matter that retains the properties of an element, and its composition—6 protons, 6 neutrons, and 6 electrons in the case of carbon—forms the *backbone of organic life*. Carbon is the fundamental element that allows for the complex chains of molecules which constitute living organisms, including proteins, DNA, and the very cells that make up human beings. In this sense, *from atom came all life*, and from carbon came the biochemical processes that animate the world. Carbon's versatile chemistry, which allows it to form stable bonds with many other elements, makes it the *cornerstone of biological molecules*, thus directly tying it to the *creation of man*.

In this context, the atom's structure with *6 protons*, *6 neutrons*, and *6 electrons* can be viewed as a *symbolic representation of man's physical nature*. Just as carbon forms the basis for all life, the numerical identity of this element in its atomic form resonates with the biblical description of man, whose creation was also a *physical act by divine will*. In the *Book of Genesis*, Adam was formed from the dust of the earth, which, in its most fundamental form, consists of atoms—the building blocks of all matter. The act of creation, whether viewed through a religious or scientific lens, is bound to the fabric of the universe, which is itself composed of these atomic structures.

THE BIBLICAL CONNECTION: NUMBER OF MAN AND ATOMIC STRUCTURE

The *Book of Revelation*, Chapter 13, verse 18, declares: *"Here is wisdom. Let him which hath understanding count the number of the beast: for it is the number of a man; and his number is Six hundred threescore and six."* The number 666, often associated with

THE INFANCY OV THOUGHT

eschatological warnings, can also be interpreted more symbolically when aligned with the concept of *"From Atom, Came Adam."* In this interpretation, the number 666, which refers to man, becomes a *numerical echo of the carbon atom*—the element with an atomic number of 6, and whose composition (6 protons, 6 neutrons, and 6 electrons) is fundamental to the construction of man and all life.

This connection emphasises the idea that man, in his physical form, is an *embodiment of the universe's most fundamental principles*. The *carbon atom* serves as the elemental basis not only for human life but for the entire biological world, and thus, the number 666 can be understood as representing the *physical essence* of humanity. Just as the atom is *the building block of the cosmos*, man, too, is a microcosm of the universe, formed from the same *celestial dust* and governed by the same physical laws. In this interpretation, the biblical reference to 666 symbolises the *material nature of humanity*, created from the elemental particles that emerged from the Big Bang and have been shaped over billions of years through the processes of *stellar nucleosynthesis* and *chemical evolution*.

THE ATOM AND THE MACHINATIONS OF THE UNIVERSE

The universe is comprised entirely of atoms, and the *machinations of the cosmos*—from the birth of stars to the formation of planets—are fundamentally driven by the interactions between these atoms. *Gravity, electromagnetism*, and the *nuclear forces* which govern atomic stability are the forces that shape the cosmos, just as they shape the human body at a molecular level. The *6 protons, 6 neutrons*, and *6 electrons* of the carbon atom make it the *element of life*, enabling the formation of complex macromolecules such as *proteins*, *lipids*, and *nucleic acids*, that are essential for the structure and function of all living cells.

The fact that the *human body* is composed of trillions of atoms, predominantly carbon, oxygen, nitrogen, and hydrogen, illustrates the *direct connection between the atom and life* itself. Each atom, with its precise configuration of protons, neutrons, and electrons, carries with it the potential to form the *complex biochemical structures* which allow life to exist. The same atomic structures that comprise the stars and planets also form the cells and tissues of living organisms, tying man to the *larger machinations of the universe* in a *profoundly interconnected web of existence*.

In this framework, the *carbon atom* becomes not just the foundation of biological life but a symbol of the *cosmic continuity* that links all matter in the universe, from the smallest subatomic particles to the vast galaxies. The processes that govern the interactions of atoms—*chemical bonding, nuclear fusion*, and *electromagnetic forces*—are the same forces that have shaped the evolution of life on Earth. The *laws of physics* that govern the movement of planets, the formation of stars, and the expansion of the universe are the same laws which govern the *molecular interactions* within a human cell.

CAUSAL REASONING FOR MAN'S CREATION

From a scientific perspective, the emergence of humanity can be seen as a direct consequence of the *chemical and physical laws* which govern the universe.

THE INFANCY OV THOUGHT

Abiogenesis, the process by which life arose from non-living matter, was driven by the *interactions of atoms* and molecules in the primordial environment of the Earth. Over billions of years, these interactions led to the formation of increasingly complex organic compounds, eventually culminating in the emergence of self-replicating molecules and the first living cells.

Once life began, *evolution*—driven by the processes of mutation, natural selection, and genetic variation—led to the diversification of species, ultimately giving rise to *Homo sapiens*. The evolution of humans was not an isolated event but the result of billions of years of *cosmic evolution*, from the formation of the first atoms to the emergence of complex multicellular organisms. The *creation of man*, therefore, can be understood as the culmination of a *cosmic process* which began with the formation of the *atom*, and through the interplay of physical, chemical, and biological forces, eventually produced a being capable of *self-awareness*, *consciousness*, and *introspection*.

Thus, *"From Atom, Came Adam"* is more than a metaphor; it is a *scientific truth* rooted in the reality that all matter in the universe, including human beings, is composed of the same *elemental particles*. The atom, with its 6 protons, 6 neutrons, and 6 electrons, represents not only the physical structure of the universe but also the *spiritual essence of creation*. It is through this profound connection that we understand our place in the cosmos—not as separate from the universe, but as *integral parts of its unfolding narrative*.

THE BALANCE OF LIFE AND THE UNIVERSE

THE INFANCY OV THOUGHT

The intricate balance of the universe, and by extension life itself, is maintained through the *complex interactions of atoms and molecules*. Just as *gravitational forces* hold the planets in their orbits and *electromagnetic forces* govern the interactions of subatomic particles, the *chemical bonds* between atoms allow for the formation of the molecules that sustain life. The same *atomic structures* which form the stars and galaxies also form the DNA that encodes the genetic information of all living organisms, further solidifying the connection between *cosmic processes* and the *biological complexity* of life.

In summation, the phrase *"From Atom, Came Adam"* encapsulates the profound truth that humanity, like all life, is borne from the same elemental particles which make up the cosmos. The *atom*, with its 6 protons, 6 neutrons, and 6 electrons, serves as the *building block of life*, and its significance extends beyond the physical to the *symbolic and spiritual*. In this understanding, the *number of a man*, 666, becomes a representation of man's connection to the *atomic structure* that underlies the entire universe. Through this lens, the creation of man is not just a biblical or spiritual event, but a *cosmic inevitability*, driven by the same *forces and laws* that govern the entirety of existence.

THE INFANCY OV THOUGHT

EDER'S SEXUAL AFFIRMATION HYPOTHESIS is an intricate model of human sexual dynamics, deeply rooted in the evolutionary mechanisms of selection and societal constructs which govern the interaction between biology and social structures. The hypothesis explores the ways in which *mating selection* exerts profound influence on evolutionary fitness, wherein males are driven by a primal imperative to secure mates through the display and selection of advantageous traits. This process, which intertwines *genetics* and *sexual behaviour*, is fundamental not only to reproduction but also to the perpetuation of desirable traits that confer *fitness advantages* across generations.

The hypothesis places a significant emphasis on the evolutionary consequences of *alpha male selection*, wherein specific male traits are selected based on their perceived ability to offer *genetic robustness*, *resource control*, or *social dominance*. The competition among males to assert dominance and attract mates is not simply a product of biological imperatives but also a consequence of *societal constructs*, where power, wealth, and status often serve as proxies for *genetic fitness* in modern human populations. Eder's hypothesis pushes this notion further by examining how *sexual affirmation*—the recognition and acceptance of one's mating potential by society and potential mates—plays a crucial role in shaping not only reproductive success but also *social dynamics*.

Sexual affirmation within this framework can be seen as a *feedback loop*, wherein individuals who successfully navigate the complex landscape of sexual selection are continuously affirmed by their environment, reinforcing their position within the social hierarchy. The alpha male, in this context, is not merely a product of brute strength or dominance but is shaped by *cognitive sophistication, social intelligence*, and the ability to navigate the intricacies of human interactions. These factors, though less tangible than physical attributes, have become critical in shaping the evolutionary success of individuals in human populations.

The implications of *sexual selection* extend beyond reproduction. Sexual affirmation is tied to *identity formation*, confidence, and social status, all which feed back into an individual's ability to reproduce and secure mates. Eder's novel perspective suggests that the *psychophysiological effects* of affirmation—or lack thereof—are crucial in determining *reproductive success*. Those who receive affirmation in their social or sexual circles are likely to experience heightened *cognitive performance, greater mental resilience*, and *increased adaptability*, as their standing within social constructs reinforces positive neural feedback. Conversely, the lack of affirmation may lead to *psychological stress*, impeding the cognitive processes necessary for successful social navigation and reducing reproductive opportunities. In essence, Eder draws a complex map where *sexual dynamics*, cognitive health, and societal pressures interlock in ways that shape evolutionary trajectories.

This brings us naturally to **EDER'S REFINED REFLEXIVITY AND NEURONAL EXCITATION HYPOTHESIS**, a theoretical framework that delves into the *neurological foundations* of cognition, adaptability, and performance under pressure. At the core of this hypothesis lies the idea that *repetitive tasks*, performed under continuous stress, lead to the enhancement of *reflexivity* and *neuronal excitation*, that in turn heighten *cognitive flexibility* and *executive function*. The repetitive nature of these tasks, often perceived as stress-inducing or monotonous, paradoxically fosters *neural efficiency* and quicker retrieval of memory, allowing the brain to handle increasingly complex situations with *greater agility*.

Eder's hypothesis connects these mechanisms to *evolutionary biology*, suggesting that the human brain, much like other parts of the body, has evolved to optimise its performance under stress. Just as the *muscle fibres* of an athlete grow more efficient through repeated strain, the brain's *neuronal pathways* become more streamlined and responsive when subjected to *intellectual or emotional stress*. This process of *neuronal plasticity*, wherein the brain adapts to meet the demands placed upon it, leads to enhanced *problem-solving capabilities*, *memory recall*, and *reflexive responses*. Continuous stress, as described by Eder, induces a state of heightened *neuronal excitation*, in that neurons fire more efficiently and in synchrony, creating *stronger synaptic connections* and thus facilitating *quicker cognitive processing*.

This *refined reflexivity* is not merely a byproduct of stress, but an adaptive response deeply rooted in *human evolution*. In survival scenarios, the ability to *improvise quickly* and react to changing environmental stimuli is paramount, and those with heightened reflexivity have a clear *evolutionary advantage*. Eder's hypothesis builds upon this evolutionary framework by suggesting that modern cognitive demands—whether they be *intellectual tasks* or *social interactions*—mimic these survival pressures, thereby honing the brain's ability to handle stress through enhanced *neuronal activity*.

This enhanced reflexivity is not an isolated phenomenon but rather connects to a larger *evolutionary paradigm* through **EDER'S ATMOSPHERIC CONDITIONING HYPOTHESIS**. Here, the focus shifts to the role of *environmental conditions*—particularly climate, geography, and atmospheric composition—in shaping *cognitive abilities* and *brain development* across human populations. The hypothesis posits that different *regional climates* have historically influenced the development of cognitive traits by imposing unique environmental challenges, thus driving *regional variation* in *intelligence* and *cognitive function*.

In regions where climatic conditions were harsh—such as extreme cold or arid deserts—survival required heightened *problem-solving skills*, adaptability, and *innovative thinking*. Eder argues that these *atmospheric pressures* acted as selective forces on early human populations, favouring traits which enabled *cognitive flexibility*, long-term planning, and *resourcefulness*. Over generations, these traits became more pronounced in populations that faced consistent *environmental adversity*, leading to *regional disparities* in cognitive development. The *brain's neuroplasticity*, combined with the *selective pressures of survival*, fostered *adaptive intelligence* tailored to specific environmental contexts.

Atmospheric conditioning thus becomes a vital evolutionary force, one that shapes not only *physical traits* but also *mental faculties*. The interplay between *climate* and *cognitive evolution* suggests that intelligence is not a fixed or uniform trait but one which has evolved in response to *regional conditions*, leading to the *cognitive diversity* seen in human populations today. This hypothesis ties into broader discussions of how *environmental factors*—including not just climate but also *geography* and *altitude*—may have influenced *neural development* over millennia, driving both the *genetic variation* and the *cultural diversity* of modern humans.

From a broader biological perspective, these mechanisms are closely intertwined with the processes of **BIOLOGICAL REPRODUCTION, MUTATION, AND RESTRUCTURING**. The role of reproduction as the vehicle through which *genetic material* is passed down across generations is foundational to *Darwinian principles* of selection and adaptation. Eder's expansion on these principles brings forth a sophisticated analysis of how *mutations*—which occur during the process of reproduction—are not merely random but, over time, are shaped by *environmental factors, social dynamics*, and *sexual selection*.

Mutations, that are often triggered by *replication errors* in DNA, introduce *novel genetic variations* into a population. Most mutations may be neutral, having no immediate effect on the organism's fitness, but occasionally, *beneficent mutations* arise, that confer advantages in survival or reproduction. These beneficial mutations are then *selected for*, gradually becoming more prevalent in a population over time. Eder's model integrates the idea that *regional atmospheric conditions*, as outlined in his Atmospheric Conditioning Hypothesis, may drive *specific mutations* to arise more frequently in certain populations, thus promoting *genetic restructuring* in response to environmental pressures.

This restructuring is not limited to the individual level but has *evolutionary implications* at the species level. As populations adapt to their environments through selective breeding, mutations, and regional pressures, *new traits* emerge that enable better *adaptation* to local conditions. These changes, driven by the mechanisms of *natural selection*, lead to *speciation* over extended periods, particularly when populations are geographically isolated or subjected to differing selective pressures. The constant interaction between *mutation* and *environmental conditioning* leads to the gradual *evolutionary restructuring* of species, resulting in the vast *biodiversity* observed across the planet.

Thus, the interplay between sexual affirmation, reflexivity, atmospheric conditioning, and biological reproduction forms an intricate web of *evolutionary dynamics*, where each factor influences and is influenced by the others. It is through these *interconnected processes* that life evolves, adapts, and diversifies, ensuring that humanity and all life forms remain resilient in the face of environmental change.

These novel hypotheses—*Sexual Affirmation*, *Refined Reflexivity and Neuronal Excitation*, *Atmospheric Conditioning*, and *Biological Reproduction, Mutation, and Restructuring*—form an intricate, interwoven framework which explains the adaptive, evolutionary processes driving human development and species divergence. By unifying these concepts, we can see how they represent different aspects of a singular, overarching theory of human evolution and adaptation, rooted in both *biological imperatives* and *environmental pressures*. Together, these hypotheses create a dynamic model which explains the complexity of human evolution in response to physical, environmental, and social stimuli.

At the core of this unified theory is the idea that *evolutionary success* is not determined by a single factor but is the result of multiple, interacting influences. *Sexual affirmation*, for instance, acts as a *social feedback mechanism*, influencing reproductive success based on traits that are both biologically advantageous and socially reinforced. This mechanism, however, does not operate in isolation. It is deeply intertwined with the *neurological and cognitive advancements* driven by *reflexivity and neuronal excitation*, where continuous stress and repetition push cognitive performance to new heights, allowing individuals to better adapt to complex social environments and competitive pressures.

The evolutionary advantage conferred by *sexual affirmation*—whether in the form of physical traits, social status, or cognitive abilities—becomes magnified through the process of *neuronal excitation*. Continuous cognitive stimulation, especially in high-pressure or competitive environments, not only enhances *reflexive responses* but also sharpens the *mental acuity* necessary for navigating social hierarchies. Individuals who excel under stress and repetitive challenges are more likely to assert *dominance* within these hierarchies, further reinforcing their *mating potential*. Thus, sexual affirmation becomes both a cause and a consequence of enhanced cognitive ability, creating a *self-reinforcing cycle* where *social validation* and *neurological enhancement* fuel one another.

THE INFANCY OV THOUGHT

The environmental context in which these processes occur is equally crucial. The *Atmospheric Conditioning Hypothesis* suggests that regional environmental pressures—such as climate, altitude, and resource availability—play a pivotal role in shaping both *cognitive development* and *genetic mutation*. Different atmospheric conditions exert unique *selective pressures*, encouraging the development of cognitive traits and physical adaptations best suited to the challenges posed by that environment. For instance, populations in colder, more resource-scarce regions might evolve greater *long-term planning abilities* and *problem-solving skills*, whereas populations in tropical environments might develop enhanced *social intelligence* and *adaptability* to rapidly changing environmental conditions.

This environmental *shaping of cognition* and *physiology* feeds directly into the processes of *biological reproduction and mutation*. As populations adapt to their regional conditions, *mutations* arise that confer specific advantages in those environments. These mutations, whether beneficial or neutral in isolation, are selected for over time, gradually *reshaping the genetic landscape* of the population. *Reproductive success*, driven in part by sexual affirmation, ensures that these advantageous mutations are passed down, leading to the *evolutionary restructuring* of entire populations. The interaction between *environmental pressures* and *biological reproduction* thus becomes a powerful force driving *speciation* and *biodiversity*.

These processes, particularly as they pertain to *human evolution* and *environmental adaptation*, are integral to the hypotheses presented, offering a framework that unites the biological, environmental, and social factors driving the evolution of Homo sapiens. By integrating these principles alongside the previously discussed concepts—*Sexual Affirmation*, *Refined Reflexivity*, *Atmospheric Conditioning*, and *Biological Reproduction and Mutation*—a more comprehensive model of human evolution emerges, one which accounts for the myriad forces shaping humanity's biological and cognitive development.

At the heart of this integration lies *natural selection*, the mechanism by which individuals with advantageous traits are more likely to survive and reproduce, passing those traits on to future generations. This principle, deeply rooted in *Darwinian evolution*, is expanded upon to address the complexities of *human evolution*. *Natural selection* is not a process confined solely to the animal kingdom; its influence permeates the human species, shaping not only physical traits but also *cognitive abilities* and *social behaviours*. Eder's work challenges the *biases inherent in scientific communities*, that have historically neglected to apply the principles of *variation and domestication* to Homo sapiens with the same scientific rigor used in the study of other species. The need to scrutinise human evolution with the same lens applied to other organisms becomes clear, especially when examining how *natural selection*, driven by *phenotypic variation*, shapes the traits which ensure survival and reproductive success.

Phenotypic variation, the observable characteristics of organisms that result from the interaction between their genetic makeup and environmental factors, plays a critical role in the process of *natural selection*. Within human populations, *selection by similarity and usefulness*—traits which confer social or reproductive advantages—becomes a key determinant of evolutionary success. Just as Darwin observed in his *Variations of Domestication*, where certain traits are selectively bred for their utility or aesthetic appeal, humans have unconsciously applied similar principles within social contexts, favouring individuals who exhibit traits that align with the demands of their environment or society. These traits, whether physical, cognitive, or social, become *adaptive advantages*, reinforced through generations by both biological reproduction and social affirmation.

This selection for advantageous traits ties directly into the previously discussed *Sexual Affirmation Hypothesis*, where traits that increase an individual's standing within a social hierarchy—whether through physical appearance, cognitive performance, or social intelligence—are *affirmed* and thereby reinforced in the evolutionary process. The *interplay between natural selection* and *sexual affirmation* creates a dynamic feedback loop, where those who are affirmed by their peers or potential mates are more likely to pass on their genes, further embedding those traits within the population.

The role of *mutation* in this evolutionary framework cannot be overlooked. *Mutations*, the random changes in an organism's genetic material, introduce *new variations* into the gene pool, which can then be acted upon by natural selection. While many mutations may be neutral or even deleterious, *beneficent mutations*—those that confer a survival or reproductive advantage—are key drivers of evolutionary change. In the case of *human evolution*, these mutations often manifest in both physical traits and cognitive abilities, shaping the course of *speciation* and the *divergence of human populations*.

Pigmentation, for example, serves as a profound illustration of how *Darwinian mutation* and *environmental adaptation* work in tandem. As discussed in the examples of *American Shorthair* and *Egyptian Siamese cats*, mutations that affect *pigmentation* are influenced by environmental factors such as *sun exposure* and *climate*. These adaptive traits, which change based on environmental needs, provide a parallel for understanding human pigmentation, where populations evolved varying levels of melanin to adapt to different levels of UV radiation. This *adaptive pigmentation* not only protects against harmful UV rays but also plays a role in *social and sexual selection*, with certain skin tones being historically valued or stigmatised based on cultural and environmental contexts.

The principle of *Darwinian mutation*, when applied to human populations, also underscores the importance of *environmental pressures* in shaping not just physical traits but *cognitive evolution*. As *atmospheric conditions* and *regional climates* shift, mutations that confer *cognitive advantages*—such as improved memory, reflexivity, or problem-solving abilities—become increasingly valuable. These mutations, shaped by *environmental conditioning*, enhance the brain's capacity to adapt to new challenges, leading to the evolution of more sophisticated *cognitive abilities* over generations.

In this context, the *Refined Reflexivity and Neuronal Excitation Hypothesis* aligns closely with *natural selection* and *mutation*. Continuous stress and repetition, that drive *neuronal efficiency*, enhance the *cognitive flexibility* required for survival in challenging environments. Individuals who can rapidly adapt to *stressful stimuli*—whether through quicker memory retrieval, heightened reflexivity, or greater adaptability—are more likely to survive and reproduce, reinforcing these cognitive traits within the population. The *mutations* that enhance *neuronal excitation* and *cognitive performance* become increasingly prevalent, particularly in populations subjected to *environmental and social pressures* that demand heightened intellectual agility.

Moreover, these processes of *natural selection* and *mutation* contribute to the larger phenomenon of *speciation*. As populations become geographically or socially isolated, the *genetic variations* that arise through mutation and selection accumulate, leading to the *divergence of species* over time. This process is not limited to physical traits but extends to *cognitive abilities* and *social behaviours*, creating distinct populations which are adapted to their unique environmental and social contexts. Over long periods, these *genetic divergences* can become so pronounced that interbreeding between populations becomes less likely, marking the emergence of new species. This process, known as *allopatric speciation*, is driven by the same forces of *mutation, natural selection*, and *environmental adaptation* that shape all life on Earth.

THE INFANCY OV THOUGHT

Eder's **ATMOSPHERIC CONDITIONING HYPOTHESIS** further refines this understanding by exploring how *regional variations in climate and atmosphere* impact the *evolution of cognition*. Different climates impose different selective pressures on populations, leading to the development of *regionally specific traits*. In colder climates, for example, where *resource scarcity* and *harsh living conditions* demand greater *strategic foresight* and *problem-solving skills*, mutations that enhance *cognitive abilities* related to these challenges are selected for. Over generations, these cognitive traits become more pronounced, contributing to the *regional differences* in intelligence and problem-solving abilities observed across human populations today. These cognitive advantages, once solidified within a population, are further reinforced by *sexual selection* and *social affirmation*, ensuring that individuals who excel in these areas are more likely to reproduce and pass on their advantageous traits.

By integrating these concepts, it becomes clear that *human evolution* is the product of a *complex interplay* between *biological mutation, natural selection, environmental conditioning*, and *social dynamics*. The selective pressures that shape human populations are not confined to the physical world but extend to the cognitive and social realms, where *sexual affirmation, cognitive performance*, and *environmental adaptability* all play critical roles in determining evolutionary success.

This unified framework emphasises the *interdependence of these forces*, where each factor reinforces and is reinforced by the others. *Mutations*, whether in the form of *physical adaptations* like pigmentation or *cognitive enhancements* like improved reflexivity, provide the raw material for evolution. *Natural selection*, driven by environmental and social pressures, acts upon these mutations, favouring those that confer an advantage in survival or reproduction. *Sexual affirmation* ensures that individuals who exhibit these advantageous traits are more likely to reproduce, embedding these traits within the population. Meanwhile, *environmental conditioning* shapes the direction of evolution, determining which traits are most advantageous in varying contexts, and thus driving the divergence of populations into distinct species.

The *coevolutionary relationship* between these factors becomes clearer when we consider the role of *feedback loops* in human evolution. Sexual affirmation, for instance, is influenced by the neurological enhancements driven by reflexivity and cognitive performance. Those individuals who excel in stressful, high-pressure environments are more likely to secure *social and reproductive affirmation*, which in turn reinforces their cognitive and social dominance. This process is further shaped by the environmental context, where the *unique atmospheric conditions* of a region determine the types of *cognitive traits* and *physical adaptations* that are most advantageous.

For example, in a region where *scarcity of resources* demands *strategic foresight* and *resource conservation*, individuals who display traits associated with *long-term planning* and *problem-solving* are more likely to succeed. Their success, in both *reproductive and social contexts*, reinforces these traits within the population, leading to the *proliferation of cognitive abilities* best suited to that environment. At the same time, environmental pressures may also lead to *mutations* that enhance these abilities at a genetic level, creating a population that is increasingly adapted to its regional conditions.

This cyclical relationship is essential to understanding the *holistic nature of evolution*. Sexual affirmation is not merely a product of social structures but is deeply rooted in *biological imperatives* that drive cognitive and physical adaptation. Reflexivity and neuronal excitation, in turn, are not isolated neurological phenomena but are influenced by the *competitive and environmental pressures* that shape human behaviour. Similarly, atmospheric conditioning is not a passive influence but an active force in determining which *traits* and *mutations* are most likely to be selected for within a population.

Synaptic plasticity, the brain's ability to change and reorganise synaptic connections in response to experience, is a cornerstone of *cognitive evolution*. In the brain, *synapses* act as communication points between neurons, where information is transmitted via neurotransmitters. When an organism is exposed to new experiences, stimuli, or challenges, these synapses can strengthen or weaken depending on the nature of the experience. This *plasticity* allows the brain to refine its neural circuits, making it more efficient at processing the information which is critical for *survival*. In the context of evolutionary theory, synaptic plasticity is not merely a mechanism for learning but a vital process that underpins an organism's ability to adapt to *environmental changes*.

As environments shift, those individuals capable of *learning from past experiences* and adapting their behaviours accordingly are more likely to survive. This is where *neurogenesis*, the birth of new neurons, plays a pivotal role. *Neurogenesis* primarily occurs in the *hippocampus*, a brain region associated with *memory formation* and is critical for the development of new *neural pathways* that support memory and learning. The creation of new neurons ensures that the brain can not only retain critical survival information but also adapt to *novel situations* by reorganizing its neural networks. This process is tightly linked to *memory consolidation*, which is the process by which *short-term memories* are converted into *long-term storage*.

Memory consolidation is essential for the retention of survival strategies, as organisms must remember previous dangers, successes, and environmental cues to navigate future challenges. The ability to consolidate and store memories effectively would have conferred significant evolutionary advantages, allowing early humans to pass down essential survival knowledge within social groups. The retention of *hunting techniques*, *foraging patterns*, and *social hierarchies* through memory was not just an individual benefit but a *collective* one, promoting the survival of early hominin groups through *shared knowledge*. This idea directly ties to *Eder's Atmospheric Conditioning Hypothesis*, where environmental pressures shape *cognitive development* over time, pushing populations to refine their *neuroplasticity* and *memory systems* to better adapt to shifting climates and resources.

The *development of consciousness* is another critical factor which must be examined within the evolutionary framework. *Consciousness*, as both an emergent property of brain complexity and a survival mechanism, allowed early humans to develop *self-awareness*, *decision-making abilities*, and most importantly, *social cohesion*. This development cannot be seen in isolation but must be understood as deeply intertwined with *social utility*—the capacity of individuals to cooperate, form *social bonds*, and create *hierarchical structures* within groups. The document explores how the development of *social traits*, such as *empathy, familial loyalty*, and *cooperation*, was essential for group cohesion and survival. Social utility is thus an evolutionary adaptation, where the ability to *work within a group, share resources*, and *protect one another* allowed early hominins to thrive in hostile environments where isolation would have meant certain death.

Empathy emerges as a *neurochemical adaptation* that facilitated *group survival*. Empathy allows individuals to understand and respond to the needs and emotions of others, strengthening the social fabric of early human communities. The *neurochemical basis* of empathy—primarily driven by neurotransmitters like *oxytocin* and *dopamine*—would have promoted *group cohesion*, as the act of caring for others was directly-linked to *reproductive success* and *survival*. Those who exhibited stronger empathetic tendencies were more likely to form *long-lasting bonds* with others, reinforcing their place within the group and ensuring that their genetic material was passed on to future generations.

The concept of *herd mentality* further develops this idea by exploring how *emotional development* and *group survival* were interlinked. Early human societies, much like animal herds, relied on *collective action* to defend against predators, gather resources, and rear offspring. The *evolutionary benefits* of *herd mentality* lie in the protection it offers through *numbers* and *collective intelligence.* By aligning behaviours with the group, individuals benefited from the *shared knowledge* and *strength in numbers* that herds provide. This phenomenon was not purely biological but was reinforced through *emotional bonds*, such as *love*, *empathy*, and *loyalty*. These emotional traits acted as *evolutionary mechanisms*, ensuring which individuals stayed together, cooperated, and *prioritised the group's survival* over individual desires.

The emotional development of humans is not a simple byproduct of evolution but a *core survival trait*, directly tied to *memory consolidation* and *neurogenesis*. Emotional experiences, particularly those involving *fear*, *love*, and *loyalty*, create *strong neural imprints*, that the brain consolidates into long-term memories. These memories are essential for *social learning*, as individuals learn from past experiences of group dynamics, cooperation, and *conflict resolution*. As emotional bonds strengthen within the group, these experiences become *engrained in the neural circuits* of individuals, fostering *social intelligence,* and promoting behaviours which enhance *group survival*.

Emotional affirmation plays a vital role in reinforcing these *social bonds* and ensuring *evolutionary fitness*. The neurochemical processes involved in *emotional bonding*—particularly the release of *oxytocin* during social interactions, love, and caregiving—serve to solidify *cooperation* and *trust* within groups. Emotional affirmation, in this context, is not simply a personal experience but a *biologically driven process* that supports group dynamics and *reproductive success*. Those who are emotionally affirmed within their group—whether through love, friendship, or loyalty— are more likely to form *stable social connections*, increasing their chances of survival and reproduction. These bonds act as *reinforcing mechanisms* for cooperative behaviour, creating a *positive feedback loop* where *emotional development* enhances *social utility*, and social utility further reinforces *group cohesion*.

The document also addresses how *evolutionary bonds* formed through *empathy*, *love*, and *loyalty* become self-perpetuating, as they are not only advantageous for *survival* but also for *reproductive fitness*. In early human societies, those who formed strong emotional bonds with others were more likely to ensure the survival of their offspring and kin, as these connections promoted *resource sharing*, *mutual protection*, and *group success*. Over time, the *neurochemical underpinnings* of these bonds—driven by *oxytocin*, *serotonin*, and other *neurotransmitters*—became deeply embedded in the *genetic code* of human populations, fostering the *emotional and social traits* that are essential for survival in complex social environments.

By integrating *natural selection*, *synaptic plasticity*, *neurogenesis*, and the development of consciousness and emotional bonds, a cohesive framework emerges in which *biological processes* and *social structures* work together to shape human evolution. *Cognitive development*, driven by synaptic plasticity and memory consolidation, enhances an individual's ability to *navigate complex environments* and *social interactions*, while *emotional bonds* ensure that these cognitive advantages are leveraged within *cooperative group structures*. As groups evolve, the *social traits* of empathy, love, and loyalty become as crucial to survival as physical adaptations, forming the bedrock of human *social evolution*.

THE INFANCY OV THOUGHT

In essence, human evolution is not solely a product of *biological mutation* or *environmental pressures*, but a *holistic process* where the brain's ability to adapt through *neuroplasticity*, the *social reinforcement of group behaviours*, and the *emotional development* of individuals are all essential components. The survival and reproductive success of Homo sapiens is tied to this intricate dance between *cognitive adaptability* and *social utility*, where *individual and collective success* are inextricably linked through both *biological imperatives* and *social dynamics*. The complexity of human evolution lies in the seamless integration of these processes, ensuring that humanity remains adaptive, resilient, and capable of thriving in an ever-changing world.

Through this unified framework, we can see how human evolution is shaped by a *complex interplay* of social, biological, and environmental factors. *Sexual dynamics*, cognitive development, and environmental adaptation are not separate processes but are deeply intertwined, each influencing and reinforcing the other in ways that drive the *divergence* and *diversification* of human populations. This interconnectedness helps to explain the vast *biodiversity* observed in human populations today, where regional differences in *cognitive abilities*, *physical traits*, and *social behaviours* can be traced back to the evolutionary pressures exerted by *environmental factors*, *sexual selection*, and *neural enhancement*.

The process of *speciation* itself, wherein populations diverge into distinct species due to *geographical isolation* or differing selective pressures, is also informed by these principles. Over long periods of time, as populations become increasingly adapted to their environments, the *genetic differences* between them become pronounced enough to prevent *interbreeding*, marking the emergence of new species. This divergence is driven by a combination of *environmental shifts*, *genetic mutations*, and *cognitive adaptation*, each factor playing a crucial role in the *reshaping of populations* and the creation of new species.

The integration of *synaptic plasticity*, *neurogenesis*, and *memory consolidation* with the broader mechanisms of evolution forms an intricate web of biological processes which are foundational to the cognitive and adaptive capacities of organisms. These processes—at their core—are deeply linked to survival, reproduction, and the continuous adaptation to environmental changes. In the evolution of species, particularly *Homo sapiens*, the ability of the brain to adapt to new information, store crucial survival strategies, and respond dynamically to external stimuli is what has ultimately shaped human cognitive evolution. By unifying these concepts with *social utility*, *herd mentality*, *emotional development*, and *emotional affirmation*, a deeply technical and holistic understanding of human evolution can be constructed, where both biological and social dynamics drive the progression of the species.

In culmination, the unification of these hypotheses offers an intricate, comprehensive model that encapsulates the multifaceted nature of *human evolution*. Evolution is not a linear process dictated by a single variable; rather, it is a deeply interconnected system in which *social validation*, *cognitive performance*, *environmental conditioning*, *genetic mutations*, and *emotional development* collectively shape the trajectory of human adaptation and speciation. The diversity observed across human populations—whether in terms of *cognitive abilities*, *physical traits*, or *social structures*—is a direct result of this *coevolutionary interplay*, wherein biology and environment engage in a perpetual dialogue, sculpting the evolution of Homo sapiens.

At the core of this framework lies *natural selection*, that remains the principal driving force behind the evolution of advantageous traits within populations. However, this model expands beyond the classical Darwinian framework to incorporate the influences of *sexual selection*, *social affirmation*, and *cognitive refinement*. Traits which confer survival advantages—whether through *physical prowess*, *intellectual agility*, or *emotional bonding*—are reinforced through *social affirmation*, ensuring that individuals who thrive in competitive, stressful, or social environments are more likely to reproduce and pass on their genetic material. This *feedback loop* between *social validation* and *evolutionary success* leads to the continuous refinement of traits that are most advantageous within specific environmental and social contexts.

The role of *synaptic plasticity*, *neurogenesis*, and *memory consolidation* in this process cannot be understated. The *brain's plasticity* allows for continuous *learning* and *adaptation*, ensuring that individuals can respond to new challenges and retain critical survival information. *Neurogenesis*—the generation of new neurons—underpins the *flexibility of cognitive systems*, allowing for the constant reshaping of *neural pathways* in response to environmental and social stimuli. This *plasticity* is critical in the evolution of *human intelligence*, as populations with heightened *cognitive flexibility* are more adept at problem-solving, social navigation, and survival in changing environments. Memory consolidation, on the other hand, serves as the foundation for *long-term learning*, allowing individuals to retain critical knowledge that can be shared and passed down through generations, further enhancing group survival.

Eder's Atmospheric Conditioning Hypothesis adds another dimension to this understanding by examining how *regional environmental pressures*, such as climate and altitude, shape *cognitive abilities* and *physical adaptations*. Over generations, populations exposed to harsh climates or resource-scarce environments develop unique *cognitive traits*—such as *strategic foresight* or *resilience*—which enable them to survive in those conditions. This *environmental conditioning* not only shapes physical adaptations, such as *pigmentation* or *body structure*, but also influences the evolution of *neural efficiency* and *cognitive performance*, pushing populations to refine their *mental faculties* in response to *environmental stressors*.

In parallel, the processes of *biological reproduction*, *mutation*, and *genetic restructuring* drive the gradual *divergence of species*. *Mutations*, though often random, provide the raw material for *evolutionary change*, introducing new variations into the gene pool that can be acted upon by *natural selection*. Over time, beneficial mutations become more prevalent within populations, leading to the development of new *traits* that enhance survival in specific environments. This continuous cycle of *genetic variation* and *selection* forms the basis for *speciation*, where isolated populations develop *distinct characteristics* that eventually lead to *reproductive isolation* and the emergence of new species.

THE INFANCY OV THOUGHT

Sexual affirmation plays a pivotal role in reinforcing these evolutionary changes. The *social dynamics* that determine an individual's *mating success* are influenced not only by *physical attributes* but also by *cognitive performance, emotional intelligence,* and *social cohesion*. Traits which enhance an individual's *status* within a group, whether through *intellectual leadership* or *emotional empathy*, are often selected for through *social and sexual validation*. This process ensures that individuals who excel in these areas are more likely to reproduce, embedding these traits within the genetic code of future generations.

The development of *consciousness* and *social utility* further enriches this evolutionary model. The ability to form *complex social bonds*, exhibit *empathy*, and cooperate with others is not merely a byproduct of evolution but a *core adaptive mechanism* that has driven human survival. *Herd mentality, emotional development,* and *familial loyalty* have allowed humans to survive as highly *social creatures*, ensuring that group cohesion and cooperation are prioritised. These *social traits*, reinforced by *neurochemical processes* such as the release of *oxytocin* during bonding, serve as mechanisms for promoting *group survival*, further enhancing *reproductive success*.

By uniting *biological, cognitive,* and *social processes*, this model presents a vision of *human evolution* as an ongoing, dynamic process. Evolution is not static; it is a *constantly shifting landscape*, where *internal genetic mutations* and *external environmental pressures* coalesce to drive *diversification* and *adaptation*. The coevolutionary relationship between *sexual selection, cognitive enhancement,* and *environmental adaptation* ensures that as environmental conditions continue to change, human populations remain resilient, adaptive, and capable of *diverging* in response to new challenges.

This model thus paints a portrait of human evolution as a *holistic process*, where *neural adaptability*, *genetic variation*, and *social dynamics* work in tandem to shape the course of human development. As long as *environmental conditions* evolve, and *human societies* continue to adapt, the forces of *coevolution* will drive the continuous diversification of human populations, ensuring that life on Earth remains *resilient*, *dynamic*, and *ever-evolving* in the face of future evolutionary challenges. The interdependence of *biology*, *cognition*, and *social utility* underscores the complexity of human evolution, making it a process that is as much shaped by *internal neurological mechanisms* as it is by *external environmental forces*. This interplay of forces, each influencing the other in *subtle yet profound ways*, ensures the continued adaptability and survival of humanity across generations.

CH. 4: SOCIAL UTILITY

SEXUAL SELECTION: REVISITED

SOCIAL INTERACTIVITY

HERD MENTALITY

FAMILIAL LOYALTY

THE INFANCY OV THOUGHT

SEXUAL SELECTION: REVISITED

In the intricate dance of human reproduction, the male's *thrusting* during *ejaculation* serves not merely as a mechanical or instinctual act but as a highly evolved behaviour designed for *maximising reproductive success*. This much is hypothesised in **EDER'S UTILISATION AND INTERACTIVITY HYPOTHESIS**, and from an evolutionary standpoint, this motion can be understood as a crucial *biomechanical function* aimed at ensuring that the *spermatozoa*—the microscopic carriers of genetic material—are delivered efficiently to their intended *ovarian target*. This movement, rhythmic and deliberate, has been shaped by the forces of *natural selection* to optimise the chances of *fertilisation*, ensuring that the genetic lineage continues with the greatest probability of success.

The mechanics of thrusting are finely attuned to the *anatomy of the reproductive system*, specifically tailored to the dynamics of *sperm delivery*. The *cervix*, acting as a gateway, lies at the entrance to the *uterus*, leading to the fallopian tubes where the *ova* await fertilisation. The thrusting action of the male serves to position the *ejaculate* as close to the *cervical opening* as possible, reducing the *distance* sperm must travel, thereby increasing the likelihood that they will reach the *ovum*. Evolution has favoured this motion because sperm, despite their small size, face significant obstacles in their journey through the *vaginal environment*, which is both *acidic* and *selectively hostile* to foreign agents. By thrusting, the male strategically bypasses these barriers, ensuring the *seminal fluid* is deposited deep into the reproductive tract, providing the sperm with a more direct and protected route toward the *egg*.

The *pleasure* associated with this act is far from incidental. The sensation of pleasure, itself an evolved *neurobiological* phenomenon, serves as a *reinforcing mechanism* that drives reproductive behaviour. The male orgasm, which coincides with ejaculation, is underpinned by a complex release of *neurotransmitters* and *hormones*, orchestrated by the central nervous system to ensure the male experiences *intense gratification*. This pleasure is not merely a by-product; it serves an adaptive function by *motivating sexual behaviour*, thereby increasing the frequency of copulation and, consequently, the chances of *reproductive success*.

At the *neurochemical level*, *dopamine* and *oxytocin* play crucial roles in generating the *rewarding sensations* of sexual climax. Dopamine, often referred to as the "reward molecule," is released in abundance during orgasm, reinforcing the behaviour by providing a powerful sense of *euphoria*. This neurological reward system is essential for ensuring that males, like all animals driven by the biological imperative to reproduce, seek out sexual opportunities with persistence and enthusiasm. *Oxytocin*, often called the *"bonding hormone,"* is also released in conjunction, fostering a sense of *connection* between partners and, in evolutionary terms, encouraging *pair bonding* and repeated mating opportunities, further promoting reproductive success.

The *hormonal interaction* that accompanies ejaculation is equally significant in the broader scope of *reproductive physiology*. *Testosterone*, the primary *androgen* responsible for male sexual drive, works in concert with these neurotransmitters to maintain *libido* and ensure that males remain driven to engage in sexual activity. This hormonal cocktail is part of an ancient evolutionary strategy, one which ensures that the biological impetus to *spread genetic material* remains a top priority. Pleasure, in this sense, becomes not only a motivational tool but a *biological necessity*, carefully crafted by millions of years of evolution to optimise the frequency and efficiency of reproductive behaviour.

Moreover, thrusting serves an additional function beyond simple sperm delivery. It has evolved as part of a *mating display*, signalling *fitness* and *dominance* to a partner. From an evolutionary perspective, *mating rituals* and *displays of vigour* have always been integral to *sexual selection*, wherein individuals demonstrate their *biological fitness* through *physical prowess* or *behavioural traits*. The act of thrusting, therefore, may also serve as a subconscious signal of *virility* and *genetic strength*, traits that females, across species, have been shown to favour when selecting mates. The force and frequency of thrusting can convey *endurance, strength,* and *overall health*, all of which are indicators that a male is likely to produce *strong, viable offspring*.

The *pleasure associated with sexual intercourse* further drives *mate retention* strategies in males. The release of *endorphins* and *serotonin* post-ejaculation helps cement the *emotional bond* between partners, thereby encouraging *long-term pair bonding*, a behaviour that has significant evolutionary advantages. Such bonding increases the likelihood of *parental investment*, particularly in species like humans, where the prolonged care of offspring is crucial to ensuring their *survival and success*. The pair bond, reinforced by the *neurochemical interplay* which occurs during and after sex, becomes an essential part of reproductive strategy, ensuring that males remain invested in their mates and, by extension, their progeny.

While the thrusting itself is a *physical manifestation* of reproductive necessity, the pleasure derived from it is a testament to how *neurobiological systems* have evolved in parallel to *physical mechanisms*. The evolution of *pleasure*—specifically sexual pleasure—has become an indispensable part of *human reproduction*, with the *neural circuits* that regulate pleasure and reward functioning as *evolutionary safeguards*, ensuring the continuation of the species. It is a profound example of how evolution has not only shaped the *physiology of reproduction* but also the *psychological* and *emotional frameworks* which support it.

Concluding, the *male thrust during ejaculation* is far more than a simple instinctual act. It is a *biomechanical strategy* finely tuned by evolution to maximise the chances of *reproductive success*. At the same time, the *pleasure* associated with it—mediated by a complex array of neurotransmitters and hormones—ensures that sexual behaviour is both *sought out* and *repeated*. This interplay between the *physical*, *neurochemical*, and *emotional* components of reproduction highlights the profound intricacy of human *evolutionary biology*, demonstrating how *pleasure, behaviour*, and *biological imperatives* have co-evolved to drive the *propagation of life*. The act of thrusting, while seemingly mechanical, stands as a *testament* to the deep evolutionary forces that govern *human reproduction* and the *complexity of human sexual behaviour*.

THE INFANCY OV THOUGHT

The phenomenon of *pleasure associated with reproductive behaviour* is not unique to human beings but is deeply rooted in the *evolutionary biology* of a vast array of species. In many mammals, particularly primates such as *chimpanzees*, *bonobos*, and other animals like *dogs*, *cats*, and *rodents*, the experience of pleasure during mating or sexual activity serves a similarly crucial role in driving *reproductive behaviour* and *social bonding*.

Chimpanzees, for instance, are known to engage in *masturbation* and *varied sexual behaviours* that are clearly motivated by *pleasure*, rather than mere procreation. This indicates that the *reward circuits* associated with *dopaminergic* and *oxytocinergic pathways*, responsible for the sensation of pleasure, have been *conserved across evolutionary lines*, shared by primates and other mammals alike. Pleasure in these contexts serves to *reinforce sexual activity*, ensuring that animals repeatedly engage in behaviours essential to the *continuation of their genetic line*.

In *chimpanzees* and their close relatives, *bonobos*, sexual activity is used not only for reproductive purposes but also as a *social tool*—for *conflict resolution*, *bonding*, and *status negotiations* within groups. This behaviour underscores the complex *interplay* between *neurochemically driven pleasure*, *social dynamics*, and *evolutionary success*. In bonobos especially, *sexual pleasure* is integral to the structure of their *matriarchal societies*, where frequent sexual interactions, often outside of reproductive necessity, reinforce *social cohesion* and reduce *aggression*. The *neurotransmitter cascades*, particularly involving *serotonin*, *dopamine*, and *oxytocin*, play a pivotal role in creating the pleasurable sensations that not only promote *sexual behaviour* but also facilitate *group stability*.

Dogs, another prime example, display *sexual mounting* behaviour that goes beyond procreation, often exhibited in contexts where *dominance* or *social play* are involved. The pleasure derived from such behaviour is mediated by similar neurochemical systems, such as the *endorphin release* which occurs during mounting. This action, while serving *social purposes* in some cases, also reinforces behaviours linked to reproductive success in the long term. Dogs, through their *olfactory and behavioural cues*, respond to stimuli which trigger *sexual arousal*, and the associated *pleasure* ensures that these behaviours are *repeated*, fostering *genetic transmission,* and ensuring the species' *survival*.

In the case of *cats*, their behaviours associated with *purring* and *kneading* during mating or affectionate interactions can also be linked to *pleasurable sensations*. While purring in cats is often interpreted as a signal of *contentment* or *relaxation*, it has been shown that *sexual stimulation* induces similar *pleasurable responses* in felines. This connection between *sensory stimulation* and *pleasure* reinforces the notion that many species have evolved *mechanisms* to *associate mating* and *sexual behaviours* with *rewarding neurochemical feedback*, ensuring these behaviours are continuously pursued.

The *biological imperative* for *pleasure* across the animal kingdom, much like in humans, is a crucial aspect of *evolutionary strategy*. The *pleasurable experiences* associated with sexual behaviours ensure that these acts are not only performed when reproductively necessary but are *sought after* proactively. The *evolutionary pressures* which shaped these neurochemical systems were likely focused on ensuring *frequency of mating*, thus increasing *reproductive success*. The release of *neurotransmitters* such as *dopamine, endorphins*, and *oxytocin* in conjunction with mating behaviours acts as a *biological reinforcement mechanism*, motivating animals to engage in sexual activities even outside of optimal reproductive conditions, thereby maximising the chances of *fertilisation* across cycles.

Beyond mammals, the concept of pleasure driving *reproductive behaviours* can be observed in other animals, albeit with varying degrees of complexity. In *birds*, for instance, elaborate *courtship displays,* and *mating rituals* often involve the stimulation of sensory systems, that, although harder to define in terms of subjective pleasure, still function to enhance the *likelihood of mating*. The *evolutionary advantage* of these pleasurable sensations lies in their ability to *reinforce* behaviours that ensure *successful reproduction* and *species continuity*.

The fact that pleasure in *sexual behaviour* is observed across a wide array of species highlights its *evolutionary importance* as a *universal mechanism*. The *neurobiological substrates* responsible for pleasure—specifically the *dopaminergic reward system*—have been conserved through millions of years of evolution, signifying their *critical role* in *driving reproductive success*. In this regard, pleasure is not a mere epiphenomenon but a *central evolutionary trait*, essential for ensuring which species remain motivated to engage in the behaviours necessary for *survival* and *genetic transmission*. This principle underscores the profound *interconnectedness* of *neurobiology*, *reproduction*, and *evolution*, revealing how the *intricacies of the mind* are deeply entwined with the *processes of life* itself.

In summation, the evolution of pleasure in sexual behaviour across species, from humans to primates to dogs and cats, is a testament to the *power of natural selection* in shaping *neurochemical responses* which reinforce *behaviours critical to survival*. These behaviours, driven by the release of *pleasurable neurotransmitters*, have become essential elements of reproductive strategies, ensuring that sexual activity is not only performed but is *continuously pursued* with vigour. This system of *biological reinforcement* has created a *universal pattern* wherein *pleasure*, tied to *neurotransmitter activity*, ensures that the *imperative to reproduce* is consistently met, thus preserving the species across evolutionary timescales.

THE INFANCY OV THOUGHT

EDER'S ORIGINATION OF HOMOSEXUALITY HYPOTHESIS

This hypothesis argues in the grand architecture of *human sexual selection*, the dynamics of *alpha* and *beta males* unfold with a complexity that speaks to evolutionary pressures shaping both *biological instincts* and *sociocultural structures*. The cultivation of *alpha males*—those who possess traits of *dominance*, *strength*, and *social status*—is not merely a product of individual will, but an emergent consequence of *speciation, reproductive strategy*, and *sexual selection* within a geographically and culturally influenced framework. The female's role in this matrix, driven by an evolved *selectivity* in her mate, adds layers of intricacy, as she inherently seeks out the *genetically fittest* males, capable of providing not just *genetic superiority* but also *security* and *resource stability* for offspring. This process, deeply rooted in *natural selection*, creates a *hierarchical system* where the alpha male stands at the apex, and the *beta male*, through competition, is either forced into social reconfiguration or evolutionary marginalisation.

The paradigm begins with the *biological foundations* of sexual selection. The concept of *survival of the fittest*, as posited by Darwin, extends beyond physical survival to encompass the *reproductive arena*, where the male's ability to *secure mates* becomes a direct measure of his *evolutionary fitness*. The male thrust to *secure genetic continuity* is an intrinsic drive, compelled by *hormonal signals* and *neurochemical feedback*. Yet, it is in the subtle nuances of *societal evolution*, cultural locality, and the geographically contingent patterns of human socialisation, that the classification of *alpha* and *beta* males gains its significance. In certain societies, the *wealth* and *status* a male accumulates may elevate his reproductive desirability, even if his *physical traits* do not align with traditional evolutionary markers of *strength* or *virility*. This evolution of *social norms* can distort the *natural order* of sexual selection, leading to an imbalance where *beta males*, artificially elevated by external factors such as *economic power*, assume roles which are evolutionarily incongruent with their biological dispositions.

The *female's selective process* is more intricate, as it is tied not only to physical attraction but also to *psychophysiological factors*. While she may inherently desire the *alpha traits*—dominance, physicality, and genetic robustness—the modern female must navigate *socioeconomic conditions*, weighing a potential mate's ability to provide *resources* and *stability*. Thus, the *sexual marketplace* is not dictated solely by *biological imperatives*, but by a *confluence of sociocultural expectations* and *economic realities*. The female's selectivity, that traditionally ensured the propagation of *strong genetic material*, now operates within a framework of *expanded criteria* which include *wealth*, *social class*, and *personal fulfilment*. This expands the spectrum of *mate selection*, but at the cost of the *naturalistic precision* that once governed *species reproduction*.

Consequently, in such *modern environments*, a paradox arises. While *alpha males* continue to engage in *multiple mating opportunities*, ensuring the *dispersal of their superior genetic material*, beta males—who traditionally would have faced evolutionary pressure to improve or adapt—find *alternative means of ascension* through *acquired wealth* or *social manipulation*. The beta male may thus *circumvent* the natural evolutionary challenges by gaining *social status* in a manner disconnected from the *biological requirements* of *natural selection*. This unbalances the gene pool, as traits once deemed unfit for reproduction are now propagated, potentially weakening the *genetic strength* of future populations.

Wherever, this temporary *anomaly in selection* is self-correcting. Over generations, the *selective pressures* will once again *assert dominance*, as *natural selection* reasserts its hold, phasing out the *weaker genetic lines* through *competitive exclusion* and *differential survival*. The evolutionary principle remains unbroken: those who cannot adapt to the *demands of survival*, whether through *physical prowess* or *genetic viability*, will ultimately be removed from the *reproductive equation*. *Alpha males*, continuing to thrive through *natural selection*, will secure the genetic future, while *betas* who rely solely on *artificial constructs* will find their *genetic lines extinguished* over time.

This leads to another complex dimension within the modern social structure—the influence of *female competition*. While *women are inherently more desirable* in reproductive terms due to the limited nature of their fertility compared to males, they too are subject to *competitive pressures* within their own gender. In environments where *alpha males* are scarce, *females compete* not just for physical attractiveness but for the *social status* and *resources* that increase their desirability in the eyes of *high-status males*. However, unlike male competition, that often culminates in direct physical or social confrontation, female competition may manifest in more *subtle forms—social signalling*,

198

manipulation, and the cultivation of traits which appeal to the *psychophysiological desires* of potential mates.

As these *sexual dynamics* play out, another layer emerges—the issue of *physicality* and *physique*. In a society where *obesity* and *ageing* have become critical factors in mate selection, the *decline in physical fitness* has led to *psychophysiological challenges*. Males, traditionally wired to select for *fertility cues*, such as *youth* and *physical health*, find themselves in a landscape where those cues are often obscured or distorted by *modern lifestyle factors*. *Female obesity* and other *health detriments* create an *evolutionary mismatch*, wherein the *selection pressures* which once favoured *physiologically optimal mates* are compromised. This adds to the *scarcity* of desirable partners on both sides of the spectrum, exacerbating the *beta-alpha divide* and further distorting the natural course of *reproductive selection*.

The hypothesis regarding the evolutionary origins and development of homosexuality, ties directly into the adaptive behaviours observed in Beta males. These individuals, unable to compete effectively with Alpha males in terms of physical dominance, reproductive success, and social status, are hypothesised to have undergone adaptive changes that allowed for their survival and integration within societal structures. This hypothesis suggests that homosexuality, as well as other non-confrontational and effeminate behaviours in Beta males, arose as a form of adaptation necessitated by their environment and social constraints.

EVOLUTIONARY MECHANISMS IN BETA MALES

THE INFANCY OV THOUGHT

Beta males, characterised by their passive-aggressive tendencies and avoidance of direct conflict, found themselves in a position where competing with Alpha males would result in failure. The Alpha males, by virtue of their physical strength, leadership qualities, and reproductive success, dominated the social hierarchy. Faced with the inevitability of this dominance, Beta males adapted by developing alternative strategies for survival and reproduction. This adaptation was not one of choice but was driven by the forces of evolution, shaping behaviours and social roles that were suited to their circumstances.

Through this process of adaptation, Beta males are thought to have developed traits that were less confrontational, more cooperative, and in some instances, effeminate. This divide between Alphas and Betas became evident through observation, with Betas avoiding direct confrontation and challenge to the status quo. Their passivity, far from being a sign of weakness, represented a survival mechanism in which they could thrive within their environment by adopting roles that did not directly challenge the dominance of the Alphas.

THE EVOLUTIONARY DEVELOPMENT OF HOMOSEXUALITY

In this context, the evolutionary roots of homosexuality can be viewed as a byproduct of these adaptive mechanisms. Beta males, unable to compete directly with Alpha males for reproductive success, are hypothesised to have developed same-sex behaviours as a form of social bonding, integration, or even as an alternative reproductive strategy. Homosexuality, within this framework, is not viewed as a deviation from natural selection but rather as an expression of it—a means by which Betas could survive and find a role within the broader social and evolutionary landscape.

The hypothesis suggests that, in the absence of reproductive success through traditional means, Beta males may have evolved behaviours that fostered cooperation, empathy, and social bonding. These traits, often viewed as effeminate, allowed Betas to navigate the social hierarchies dominated by Alphas without resorting to direct confrontation. Homosexual behaviours, in this sense, may have developed as a form of social alliance, where Beta males could form strong bonds with both Alphas and other Betas, ensuring their survival within the group.

SOCIAL AND ENVIRONMENTAL INFLUENCES

Social and environmental factors are believed to have played a significant role in shaping the adaptive behaviours of Beta males. In early human societies, physical strength and dominance were paramount for survival and reproductive success, qualities that Alpha males possessed in abundance. Betas, however, found their niche through cooperation and emotional intelligence. Their ability to avoid direct competition with Alphas allowed them to survive and thrive, despite their lesser physical capabilities.

In modern environments, where social norms and cultural pressures have shifted, the acceptance and even celebration of diversity in sexual orientation and gender roles may be seen as an extension of these earlier adaptive behaviours. Homosexuality, once an adaptive behaviour for social integration and survival, has evolved into a more complex social construct shaped by contemporary cultural and social forces. However, the underlying evolutionary mechanisms remain, with homosexual behaviours continuing to serve as a means of forming social alliances and navigating social structures.

EFFEMINACY AND THE BETA MALE

Effeminacy in Beta males is a key aspect of this evolutionary hypothesis. As Beta males lacked the physical dominance and reproductive prowess of their Alpha counterparts, they are thought to have evolved traits that were more emotionally complex, cooperative, and non-threatening. These traits, that are traditionally associated with femininity, allowed Beta males to survive by avoiding direct competition with Alphas. The adoption of effeminate traits, far from being a sign of weakness, was an adaptive strategy that enabled Betas to integrate within social hierarchies and find alternative means of survival.

Effeminacy in Beta males, along with the development of homosexual behaviours, can thus be understood as an evolutionary response to their environment. Unable to compete directly with Alphas, Betas evolved traits that made them more socially adaptable and allowed them to form bonds that were crucial for their survival. This adaptation, driven by the forces of natural selection, resulted in a clear divide between the aggressive, dominant Alphas and the passive, cooperative Betas.

THE ROLE OF WILLED MOTION

THE INFANCY OV THOUGHT

In the broader framework of Infancy Ov Thought, the concept of Willed Motion as the prime catalyst of existence ties into the evolutionary mechanisms that shaped the behaviours of Beta males. Willed Motion, as the force that initiated the first spark of creation, can also be viewed as the driving force behind the adaptive behaviours seen in Betas. Just as Willed Motion set the universe into motion, so too did it guide the evolutionary processes that shaped the survival strategies of Beta males.

Betas, like all beings, are driven by the force of survival and existence. Their adaptive behaviours, including the development of homosexual tendencies and effeminate traits, are expressions of this larger evolutionary drive. Homosexuality, within this framework, is not an aberration but an example of how Willed Motion manifests through the diverse and complex behaviours that ensure the survival and persistence of life.

HOMOSEXUALITY AS AN EVOLUTIONARY EXPRESSION

Homosexuality in Beta males is hypothesised to have emerged as a natural consequence of their adaptive strategies. In the absence of physical dominance, Beta males relied on emotional intelligence, cooperation, and non-confrontational behaviours to survive. These behaviours, including homosexual bonding, served as alternative strategies for social integration and survival. Homosexuality, within this evolutionary framework, is seen as an expression of the Beta male's Willed Motion—the drive to survive and persist in an environment dominated by Alphas.

In essence, homosexuality represents an alternative form of social bonding and reproduction, where Beta males found their place within the social hierarchy through emotional and social alliances rather than physical dominance. This adaptation allowed Betas to thrive and contribute to the broader genetic and social diversity of the species. The evolutionary development of homosexuality in Beta males, as posited in *Infancy Ov Thought*, represents a complex adaptive strategy shaped by the forces of natural selection and Willed Motion. Unable to compete directly with Alpha males, Betas evolved behaviours that allowed them to survive and integrate within social hierarchies. Homosexuality, far from being an anomaly, is viewed as an evolutionary expression of these adaptive behaviours, ensuring the survival and persistence of Beta males within the broader context of human evolution. Through this lens, homosexuality is understood as part of the diversity and complexity that defines the human species, shaped by evolutionary mechanisms that continue to influence behaviour and social structures.

THE INFANCY OV THOUGHT

EDER'S 'INCLUSIONARY SELECTION HYPOTHESIS,'

Within this context, another point of divergence emerges: the phenomenon of *homosexuality*. From an evolutionary perspective, *homosexuality* does not contribute directly to *reproductive success*, yet it persists across species. The suggestion that *homosexuality* may arise from *psychological trauma*, *emotional abuse*, or *isolation* points to a *non-genetic* explanation—one which is *socially and psychologically contingent*. The same-sex attraction, in this hypothesis, is seen not as an *innate biological drive* but as a response to *emotional deprivation*, where the individual seeks *comfort* and *security* through relationships which mirror their own *emotional state*. This theory posits that such behaviours emerge not from a *biological imperative* but from a *psychosocial necessity*, a coping mechanism in the face of *emotional hardship*.

The evolutionary mechanisms that underlie human sexuality, particularly in the context of homosexuality, demonstrate a profound complexity when examined through the lens of natural selection, biology, and sociocultural adaptation. The cycle of life and reproduction, driven by the desire for genetic survival and continuation of species, shapes the behaviours and selections of individuals within a population. This process is visible across species, where evolutionary pressures encourage individuals to seek out traits that enhance their reproductive success. Within this evolutionary context, both heterosexual and homosexual behaviours may be seen as part of a broader adaptive landscape.

In the case of heterosexual relationships, young women, typically as early as age fifteen, begin to exhibit behaviours aligned with natural selection's preference for the fittest partners—those with the strongest and most advantageous genetic traits. At this stage of reproductive maturity, females are unconsciously driven to seek out males who exhibit signs of evolutionary fitness. These traits, from a biological standpoint, include strength, leadership qualities, and intellectual capacity, all of which signal a male's ability to provide genetic material most likely to succeed in offspring survival and reproduction.

From a purely evolutionary perspective, the traits women seek in men serve specific biological functions: reproductive success, sperm survival, and overall fitness. This process of selection ensures that gene pools remain healthy, avoiding weaknesses or diseases that could jeopardise future generations. However, the landscape of sexual attraction is far more intricate, with variations in male and female roles revealing layers of selection pressures that go beyond the straightforward binary reproductive narrative.

SEXUAL DYSPHORIA AND EUPHORIA IN EVOLUTIONARY CONTEXT

The distinction between Alpha and Beta males in terms of evolutionary roles sheds light on the broader mechanisms of selection and genetic perpetuation. Alpha males, characterised by larger reproductive organs and a more dominant social presence, tend to exhibit traits that make them more desirable for breeding from a natural selection viewpoint. Their physical dominance, larger sexual organs, and increased fertility rates compensate for a perceived lack of intellectual development, placing them in the role of primary breeders within the population. In contrast,

Beta males, while possessing larger brains, focus on intellectual pursuits and social cooperation, filling roles that contribute to the broader survival of the species through different, albeit vital, means.

Natural selection favours both Alpha and Beta males in their respective capacities. While Alphas are evolutionarily designed to ensure the propagation of their genes through physical traits, Betas contribute to the intellectual and social development of the population. This dichotomy highlights the inherent complexity in evolutionary strategies, as survival is not solely dependent on reproductive success but also on the development of societal structures, problem-solving, and adaptive intelligence.

NATURAL SELECTION'S ROLE IN HOMOSEXUALITY

Homosexuality, within the framework of evolution, presents an intriguing case. While homosexual relationships do not result in direct reproduction, they may serve an adaptive function within social groups. Homosexual behaviour has been documented in many species, including dogs and other mammals. Although the presence of homosexual relations in animals is often used to argue for the naturalness of such behaviour, it remains speculative whether these instances offer definitive evidence for evolutionary advantages.

Homosexuality could be viewed through the lens of kin selection or social bonding, where non-reproductive individuals contribute to the well-being of their kin group, thereby enhancing the survival of shared genes. This form of altruism—where non-breeding individuals provide support within their social group—may offer an indirect evolutionary benefit by fostering stronger, more cohesive communities that can better withstand environmental pressures.

EVOLUTIONARY ADAPTATION AND ENVIRONMENTAL INFLUENCE

The survival of human species during the Ice Age and beyond is a testament to the flexibility and adaptability inherent in natural selection. Human diets, lifestyles, and even psychological states were all subject to the pressures of their environments. The harsh conditions of the Ice Age necessitated diets high in fats and proteins, and evolutionary mechanisms allowed for survival under such extreme conditions. Northern populations, that faced longer winters and harsher climates, developed traits that favoured intellectual growth and adaptability, as physical exertion was limited due to the need to conserve energy.

Conversely, populations in warmer climates, where physical exertion was crucial for survival due to heat and humidity, developed traits that favoured physical fitness and sexual activity. These environmental differences contributed to divergent evolutionary paths, with northern populations placing greater emphasis on intellectual development, while southern populations developed stronger physical traits and reproductive capacities. Such divergence illustrates the complexity of human evolution, where environmental factors, dietary habits, and social structures interplay to shape human physiology and psychology.

THE INFANCY OV THOUGHT

TIME MANAGEMENT, MULTIPLE INTELLIGENCES, AND THE EVOLUTIONARY ORIGINS OF SEXUAL BEHAVIOUR

When considering Gardner's Multiple Intelligences within this evolutionary framework, it becomes clear that human intelligence and behaviour are influenced by a range of factors, including environmental pressures, genetic makeup, and social structures. The idea of time management, for instance, is not merely a modern construct but one deeply rooted in evolutionary mechanisms. The ability to plan, manage resources, and allocate energy efficiently has been critical to the survival and success of human populations.

Homosexuality, viewed through this broader evolutionary lens, may represent one of many diverse adaptive strategies that contribute to the overall fitness of the population. While homosexual individuals may not directly reproduce, they may play crucial roles in supporting kin, maintaining social cohesion, and contributing to the intellectual and cultural development of their communities.

THE COMPLEXITY OF EVOLUTION AND THE INTERPLAY OF VARIABLES

The evolutionary mechanisms governing human sexuality and behaviour are deeply complex, involving layers of physical, scientific, philosophical, and mathematical considerations. Natural selection, while a powerful force, operates through a variety of pathways, influencing not only physical traits but also intellectual, social, and emotional development. Whether through the direct propagation of genes or through the support of kin and community, individuals contribute to the survival and evolution of their species in myriad ways.

The cyclical nature of life and reproduction, shaped by evolutionary pressures and environmental factors, demonstrates that human sexuality and behaviour are not rigid constructs but fluid, adaptive responses to the challenges of survival. Homosexuality, in this context, may be understood not as an anomaly but as one of many diverse strategies through which human beings navigate the complex interplay of genetics, environment, and social structures.

In analysing the origins and evolutionary mechanisms behind human sexuality, particularly within the framework of homosexuality, it becomes essential to consider not only biological and environmental factors but also the influence of social and psychological variables. The universality of human intelligence and consciousness—the ability to perceive, reason, and interpret—plays a pivotal role in how individuals interact with and interpret the world around them. Empirical reality, as it is experienced, is undoubtedly true; however, when societal constructs or external influences interfere, they can obscure or distort that experiential reality. In this context, human behaviour, including homosexuality, may not only be shaped by genetic or environmental factors but also heavily influenced by social conditioning and learned behaviours. Homosexuality, while observed in various species, may not be solely a product of natural selection or genetic predisposition. Human influence, learned behaviour, and social conditioning play a significant role in shaping sexual identities and behaviours. The concept of Universality Intelligence—the collective human capacity to perceive and interpret reality—

suggests that behaviours like homosexuality are complex and influenced by a variety of factors, including social norms, peer influence, and the broader cultural context.

Just as life itself is cyclical, so too are the forces that shape human behaviour. The interaction between natural selection, social conditioning, and human intelligence forms a multifaceted web of influences that guide the evolution of human sexuality. Homosexuality, in this view, is not an isolated phenomenon but part of a broader spectrum of human experience, shaped by both evolutionary mechanisms and social structures. The balance between biological imperatives and social constructs ensures that human behaviour remains dynamic, adaptive, and deeply intertwined with the complexities of existence.

UNIVERSALITY INTELLIGENCE AND THE ROLE OF HUMAN INFLUENCE

In considering the concept of Universality Intelligence, which encompasses the collective human capacity to interpret, understand, and navigate the complexities of existence, it is essential to acknowledge how both individual and societal factors shape human behaviour. All that we experience is empirically real, but the distortions of memory, perception, and societal conditioning can lead to the formation of behaviours that may not have originally aligned with natural evolutionary functions. Homosexual relations, for instance, could be viewed through the lens of human influence, rather than being solely a biological imperative. Human influence, particularly when dealing with unnatural or unobserved behaviours in nature, plays a crucial role in shaping non-reproductive sexual behaviours.

Homogeneous capacity—or the uniform potential for reproductive and social behaviours across species—is primarily driven by evolutionary autonomy. Within this natural framework, behaviours that support survival and reproduction are typically reinforced. However, human intervention introduces variables that disrupt this autonomy. For instance, in cases where children are exposed to inappropriate sexual behaviour, they lack the understanding to discern moral from immoral acts. Similarly, the behaviour of animals, such as male dogs engaging in homosexual acts, may be attributed not to an inherent evolutionary trait but rather to learned or observed behaviours facilitated by human influence.

SOCIAL CONDITIONING AND THE PERCEPTION OF HOMOSEXUALITY

Social conditioning, along with the desire to belong or fit within a societal structure, further complicates the evolutionary narrative of homosexuality. While the biological and environmental arguments for a "gay gene" remain inconclusive, social factors play an undeniable role in shaping sexual orientation. Many individuals who identify as homosexual often report that they were not subjected to traumatic experiences, such as rape, molestation, or trafficking, that would have directly influenced their sexual preferences. However, this does not eliminate the possibility that social norms and pressures contribute to the development and expression of homosexual behaviour.

THE INFANCY OV THOUGHT

Social dynamics, such as cliques, subcultures, or movements, often create environments in which individuals seek validation or a sense of belonging. In some cases, homosexuality, like other identities, becomes part of a social revolution, a means of aligning oneself with a perceived underdog or oppressed group. The normalisation of queerdom, therefore, could be seen as part of this larger social dynamic, where identity and sexual behaviour are shaped by social movements, peer influence, and the desire for inclusion or persecution.

EVOLUTIONARY ADAPTATION AND SOCIAL INFLUENCE

When viewed through the evolutionary lens, human behaviour is largely driven by the survival of the species and the perpetuation of genes. In the case of homosexuality, while it may not serve a direct reproductive function, it could still be seen as a byproduct of social structures that provide other forms of support or cohesion within a population. The kin-selection hypothesis suggests that non-reproductive individuals might contribute to the survival of their kin group by assisting with the rearing of offspring, providing resources, or maintaining social bonds.

Moreover, it is important to understand how homosexuality, in certain instances, may be driven not by biological imperatives but by the broader social norms and pressures that influence human behaviour. In cultures or societies where queerness becomes increasingly accepted or celebrated, the likelihood of individuals exploring or adopting such identities increases—not necessarily as a direct result of genetic predisposition, but through learned behaviours and socially encouraged experimentation.

THE CYCLICAL REPETITION OF LIFE AND HUMAN INTELLIGENCE

The cycle of life is both beautiful and sorrowful, as each generation seeks to contribute more than the last, though history often remains ambiguous in its judgment of such progress. Young individuals, particularly women, begin seeking genetically fit partners at early ages—evidence of natural selection's enduring influence on human mating strategies. However, the selection process today is no longer purely biological; social norms, education, and psychological influences also dictate whom individuals choose as mates. Just as evolutionary pressures guided humanity in ancient times, today's social structures and cultural revolutions influence individual choices, including sexual orientation.

Within this broader evolutionary narrative, human intelligence plays a crucial role in how individuals navigate life's complexities. The brain, acting as a vast library of knowledge, experiences, and learned behaviours, processes an enormous range of stimuli—from biological imperatives like reproduction to social constructs that shape identity. It is this very intelligence, this Universality Intelligence, that enables humans to interpret their place in the world, question social norms, and make choices that either align with or deviate from traditional evolutionary paths.

THE INFANCY OV THOUGHT

REWORKING GARDNER'S MULTIPLE INTELLIGENCES AND EVOLUTIONARY MECHANISMS

The evolutionary mechanisms behind human intelligence, particularly in relation to sexuality, can also be tied to Gardner's theory of multiple intelligences. Human intelligence is multifaceted, including logical-mathematical reasoning, social intelligence, and spatial awareness. Time management, for example, plays a crucial role in both survival and intellectual development, as individuals who can plan and allocate resources efficiently are more likely to succeed. In the case of sexual orientation, social intelligence—the ability to navigate complex social environments—may be particularly relevant, as sexual behaviour can be influenced by social dynamics, peer groups, and societal expectations.

In preclusion, the *modern sexual landscape* reveals a *multifaceted interaction* between *biological imperatives*, *social constructs*, and *psychological realities*. The *alpha-beta dynamic*, though rooted in *evolutionary biology*, is now filtered through the lens of *socioeconomic structures*, *cultural expectations*, and *psychological adaptations*, creating a complex web of *sexual selection* and *genetic propagation*. Ultimately, the forces of *natural selection* will continue to shape human reproduction, even in the face of *modern distortions*, reaffirming the *evolutionary principles* that have governed life since the dawn of existence.

The *psychosocial roots* of the phenomena described, delve into the intricate *interplay between social hierarchies*, *biological imperatives*, and *evolutionary survival strategies*. The hierarchical structure in which

THE INFANCY OV THOUGHT

alpha males dominate access to *multiple female partners*, leaving *beta males* with limited or no opportunity for mating, has profound *psychological, biochemical,* and *social consequences* which shape *male-male interactions* and their coping mechanisms. As described in *Sapolskian studies, stress* from *social ostracisation, loneliness,* and *isolation* can trigger a cascade of *neurochemical imbalances*, including *elevated cortisol levels* and *reduced dopamine activity,* that have tangible effects on behaviour, emotional state, and social bonding.

The dominance of *alpha males* in securing multiple partners results in a form of *social scarcity* where *beta males,* unable to compete for *female attention,* may face *chronic stress* similar to the socially subordinate baboons observed in *Sapolsky's research*. These males, deprived of access to reproductive opportunities, may undergo physiological changes such as *increased cortisol production* due to *chronic social stress*, which can lead to *depressive states, decreased libido,* and overall *reproductive dysfunction.* The *biological drive* for reproduction, however, does not dissipate. Instead, it forces the individual to find *alternative pathways* to fulfil *social needs* and *emotional stability,* often through *non-reproductive bonding* with other males.

The *inquisitive nature* of *beta males*, combined with *social deprivation*, could lead them to explore *same-sex interactions*, not necessarily with an initial sexual purpose, but through a desire for *emotional intimacy* and *social bonding*. As in all mammalian species, *social connection* plays a critical role in *emotional regulation, stress reduction,* and *survival strategies*. Lacking access to *heterosexual relationships, beta males* may have instinctively turned to one another, seeking *comfort, companionship*, and a way to mitigate the overwhelming *stress of isolation*. This *social interactivity* and *emotional closeness*, built out of necessity, could have laid the groundwork for *homoerotic bonds*, where *physical interactions* might follow from a desire to replicate the *closeness* and *comfort* typical in *heterosexual relationships*.

This is not to argue that *homosexual behaviour* in humans is purely adaptive or situational, but rather to suggest that *environmental pressures*, such as *social isolation* and *stress*, may have contributed to *same-sex bonding* in a historical context where *female mates* were *inaccessible* to *beta males*. The *emotional strength* found in these relationships likely stemmed from shared experiences of *rejection* and *ostracisation* by both *alpha males* and the *females* who selected them. As these bonds deepened, they may have taken on forms which mirrored the *emotional and psychological stability* found in *heterosexual pairings*, even if *reproduction* was not the direct goal.

In this framework, *biochemical factors* further influence the development of such relationships. The *neurochemical mechanisms* which reinforce *emotional bonding* in *heterosexual relationships*, particularly the roles of *oxytocin, vasopressin*, and *dopamine*, would similarly facilitate *bond formation* between *beta males*. *Oxytocin*, known for its role in *social bonding*, would strengthen emotional ties, while *dopamine*, the brain's *reward neurotransmitter*, would reinforce positive feelings associated with *intimacy* and *closeness*. Over time, the repeated release of these *neurotransmitters* could condition males to *seek* and *find emotional fulfilment* within *same-sex relationships*, solidifying bonds which, although not initially reproductive, become socially and psychologically significant.

This hypothesis also offers a potential explanation for the *societal persecution* of *LGBT individuals* in certain historical and cultural contexts. As *alpha males* maintained their *dominance* in the *sexual hierarchy*, they could have viewed *beta males' same-sex relationships* as a *threat* to their *status* or *reproductive monopoly*. This *social tension*, coupled with the need to protect *heteronormative structures* which ensured *alpha male dominance*, may have contributed to the *marginalisation* and *persecution* of *LGBT individuals*. The *stigmatisation* of these bonds could be viewed as a way for *alpha males* to *reaffirm* their *social superiority* and maintain *control* over the *reproductive gene pool*, while *beta males*, ostracised from reproductive competition, were further *marginalised* for forging *non-reproductive alliances*.

Additionally, from a *psychosocial perspective*, the *fear of abandonment* by potential female mates, combined with *ostracisation by alpha males*, would have driven *beta males* to seek *alternative sources* of *social validation* and *emotional support*. The *societal constructs* that eventually formed around *heterosexual norms* could have amplified the feelings of *insecurity* and *inadequacy* among *beta males*, pushing them toward *same-sex interactions* as a means of *mitigating psychological stress*. In this light, *same-sex relationships* may have functioned as a *coping mechanism*, providing *emotional refuge* in a world that marginalised their reproductive potential and left them with *limited paths to societal fulfilment*.

THE INFANCY OV THOUGHT

These *stress-induced pathways* toward *homosocial bonding* could also explain why *same-sex relationships* sometimes mirror the *emotional dynamics* and *commitment* seen in *heterosexual pairings*. The *psychological need* for *affirmation*, *stability*, and *belonging*, underpinned by *evolutionary pressures* to find security in numbers, would naturally lead to *deep, enduring bonds*. These relationships may provide a *mutual defence* against *psychological breakdown* in the face of *social exclusion* and could be seen as an *adaptive strategy* in environments where *reproductive opportunities* are sparse.

The persistence of *homosexual behaviour* across many species, not just humans, adds weight to this idea. In *social animals* like *bonobos* and *dolphins*, *same-sex interactions* are observed not only as *sexual behaviours* but as *mechanisms of social bonding, conflict resolution*, and *group cohesion*. These behaviours may have *evolutionary significance*, even if they are *non-reproductive*, by *enhancing group survival* and *strengthening alliances* within the social hierarchy. The *biochemical reinforcement* of *bonding hormones* such as *oxytocin* plays a central role in maintaining these relationships, whether they are *heterosexual* or *homosexual* in nature.

Evolutionarily speaking, the existence of *same-sex relationships* can be seen not as a deviation from the reproductive norm but as an *alternative strategy* for *social survival*—of a means of and to *survival*. In environments where *females* are scarce or where *beta males* face *systemic exclusion* from *reproductive opportunities*, these *relationships* serve to mitigate the profound *stress* associated with *loneliness, ostracisation*, and *social isolation*. Through *emotional bonds* which foster *stability*, *mutual protection*, and *shared experiences*, these relationships provide a *vital psychosocial buffer* in the face of *evolutionary pressures*.

In sum, this theory posits that *same-sex interactions*, particularly among *beta males*, may have arisen as an *adaptive response* to *environmental pressures*, primarily the *social dominance* of *alpha males* and the *scarcity of available female partners*. The combination of *psychological need, biochemical reinforcement*, and *evolutionary strategy* forms the foundation for understanding how these relationships, though not directly reproductive, play a crucial role in *emotional regulation, social cohesion*, and ultimately, the *survival of individuals* who might otherwise be left *isolated* and *disenfranchised* within the reproductive hierarchy. This unique *social dynamic*, while distinct from *heterosexual pairings*, shares the *neurochemical and emotional substrate* that defines human social interaction, rooted deeply in the evolutionary biology of *survival* and *connection*.

THE INFANCY OV THOUGHT

"THE HONEYMOON PHASE, IN CYCLICALITY" HYPOTHESIS

The hypothesis regarding the psychological and physiological benefits of repeated cycles of falling in love and breaking up proposes an intriguing perspective on how these cycles might contribute to emotional resilience, neurochemical adaptation, and overall well-being. Within the framework of Infancy Ov Thought, this hypothesis can be viewed as an extension of human adaptive mechanisms and survival strategies, tied to the concept of Willed Motion and the intricate dynamics of evolutionary biology.

THE EVOLUTIONARY ROOTS OF ROMANTIC CYCLES

In the context of evolutionary biology, romantic love serves as a powerful biological mechanism designed to foster social bonding, reproductive success, and emotional security. The cycle of falling in love and experiencing a break-up may not merely be a result of cultural or societal factors but rather an intrinsic evolutionary process. This process could be instrumental in honing emotional resilience and fostering psychological adaptation in response to the complexities of human relationships.

From an evolutionary standpoint, the heightened emotional experiences associated with romantic love, including the release of key neurotransmitters such as oxytocin and norepinephrine, are essential for forming strong social bonds and increasing reproductive success. However, the adaptive benefits of experiencing multiple romantic cycles—where individuals repeatedly engage in the process of forming emotional bonds and enduring their dissolution—could extend beyond immediate reproductive purposes.

NEUROBIOLOGICAL FOUNDATIONS: OXYTOCIN, NOREPINEPHRINE, AND CORTISOL

The neurobiological mechanisms involved in falling in love and breaking up are central to this hypothesis. Oxytocin, known as the "love hormone," plays a crucial role in bonding, attachment, and social connectivity. It is particularly released during moments of physical and emotional intimacy, which contribute to an individual's overall sense of trust and security. In evolutionary terms, oxytocin fosters pair-bonding and group cohesion, which are essential for survival. The repetitive experience of falling in love and engaging in emotional bonding could serve to strengthen these neurobiological pathways, enhancing one's ability to form social connections and maintain psychological stability over time.

Norepinephrine, on the other hand, is associated with the heightened arousal and excitement that accompanies the early stages of romantic relationships. This neurotransmitter is critical in enhancing attention, memory encoding, and emotional salience. Its repeated activation through successive relationships could improve cognitive functions associated with memory, motivation, and emotional engagement, thus allowing individuals to process emotional experiences more effectively and with greater depth.

Additionally, cortisol, the body's primary stress hormone, is typically elevated during periods of emotional distress, such as break-ups. However, during the honeymoon phase of romantic love, cortisol levels decrease, inducing relaxation and a sense of well-being. Repeated cycles of falling in love and breaking up could enhance the brain's ability to manage stress by regularly reducing cortisol levels and promoting recovery after emotional turmoil. Over time, this cyclical process could condition the hypothalamic-pituitary-adrenal (HPA) axis to respond more adaptively to stress, leading to improved stress resilience and psychological equilibrium.

PSYCHOLOGICAL RESILIENCE AND EMOTIONAL GROWTH THROUGH CYCLIC ROMANTIC EXPERIENCES

The hypothesis also postulates that engaging in repeated romantic cycles allows for emotional growth and psychological resilience. Each experience of falling in love, followed by the emotional challenge of a break-up, could serve as an opportunity for individuals to refine their emotional coping strategies. Over time, this repetitive exposure to emotional highs and lows enables individuals to build emotional strength, enhance self-awareness, and develop adaptive responses to future relationships.

This adaptive emotional growth aligns with evolutionary mechanisms that favour individuals capable of handling stress and social dynamics with greater efficiency. Much like Beta males, who adapted by developing non-confrontational, socially integrative behaviours in response to the dominance of Alpha males, individuals who engage in these repeated romantic cycles may be honing their emotional intelligence and resilience as a form of evolutionary adaptation. The ability to emotionally recover from a break-up and re-engage in new relationships can be viewed as a strategy that ensures emotional survival and enhances social connectivity within complex human societies.

THE ROLE OF WILLED MOTION IN ROMANTIC CYCLES

Within the broader theoretical framework of Infancy Ov Thought, the concept of Willed Motion plays a pivotal role in understanding these cyclic romantic experiences. Willed Motion, as the fundamental force behind creation, evolution, and the persistence of life, extends beyond mere physical existence to encompass the emotional and psychological realms of human experience. The act of repeatedly falling in love and experiencing break-ups could be seen as an expression of Willed Motion—an ongoing process of emotional evolution and growth driven by the forces of neurobiological and psychological adaptation.

Just as *Willed Motion* guides the evolution of the cosmos through cycles of creation, destruction, and rebirth, so too does it govern the emotional cycles that individuals experience in their romantic lives. These cycles, characterised by the highs of emotional bonding and the lows of emotional dissolution, are not merely random occurrences but are reflective of deeper evolutionary processes that shape human behaviour and emotional resilience. In this sense, the experience of falling in love and breaking up is part of the broader continuum of life, driven by the same primordial forces that guide the evolution of the universe.

THE INFANCY OV THOUGHT

THE HONEYMOON PHASE AND THE BETA-ALPHA DIVIDE

The hypothesis surrounding repeated romantic cycles also ties into the evolutionary dynamics between Alpha and Beta males. Betas, characterised by their more passive-aggressive nature and adaptive tendencies, may have developed an evolutionary strategy that involves emotional bonding and social cooperation rather than direct competition with Alphas for reproductive success. The cyclical experience of romantic relationships, particularly the emotional highs of the honeymoon phase, provides an opportunity for Betas to reinforce their social bonds and emotional resilience without directly challenging the dominant Alpha males.

In this context, the honeymoon phase of relationships, marked by heightened oxytocin and norepinephrine levels, can be seen as a biologically adaptive period that allows individuals—particularly Betas—to establish emotional connections and social alliances. These connections serve to reinforce their place within the social hierarchy, allowing them to thrive in a non-confrontational, emotionally adaptive manner.

HORMONAL ADAPTATION AND LONG-TERM WELL-BEING

The repeated reduction in cortisol levels, coupled with the consistent release of oxytocin and norepinephrine during successive romantic relationships, could lead to long-term hormonal and neurochemical adaptations. Over time, these adaptations may result in a finely-tuned stress response system, improved emotional regulation, and enhanced psychological well-being.

The ability to navigate the emotional turbulence of romantic cycles, while maintaining a balanced and resilient emotional state, reflects the adaptive nature of the human mind and body.

Furthermore, these cycles of emotional bonding and recovery could lead to long-term psychological benefits, including improved emotional intelligence, greater self-awareness, and a more robust capacity for stress management. By engaging in these romantic cycles, individuals may be better equipped to handle the emotional challenges of life, fostering emotional resilience and psychological growth in the face of adversity.

CYCLIC ROMANTIC EXPERIENCES AS AN EVOLUTIONARY STRATEGY

The hypothesis surrounding the psychological and physiological benefits of repeated cycles of falling in love and breaking up aligns with broader evolutionary theories and the concept of Willed Motion as presented in Infancy Ov Thought. By engaging in these romantic cycles, individuals may be tapping into deeper evolutionary mechanisms that foster emotional resilience, neurochemical adaptation, and long-term psychological well-being. Through the repeated release of key neurotransmitters such as oxytocin and norepinephrine, and the reduction of cortisol levels, individuals are able to navigate the emotional complexities of life with greater ease and adaptability. This process, driven by the same forces that guide the evolution of the universe, reflects the intricate interplay between emotional experience, neurobiology, and the broader forces of evolution.

CH. 5: NATURE

PSYCHOPATHY (INTRAPERSONAL AND INTROSPECTIVE LONELINESS AND ISOLATION CAUSAL TO DISORDER)

- APATHY
- HATRED
- INABILITY TO RECONCILE FEELINGS
- ABANDONMENT
- HEARTBREAK
- RAGE
- PSYCHOSIS
- SADISM

THE INFANCY OV THOUGHT

PSYCHOPATHY
SYSTEMIC, BY NATURE

The intricate interplay of *genetic predispositions* which culminate in the development of psychopathy and its associated traits—such as *apathy, hatred, inability to reconcile feelings, abandonment, heartbreak, rage, psychosis,* and *sadism*—is rooted deeply within the *biological architecture* of the brain, independent of environmental influences. These emotions and behavioural tendencies, manifesting in psychopathy, can be understood as the result of complex *neurological and genetic interactions* that shape an individual's emotional responses, cognitive processing, and social engagement.

At the genetic level, *psychopathy* can be seen as a disorder emerging from a combination of inherited *neurobiological abnormalities* that influence the development and function of key brain regions involved in *empathy, emotional regulation*, and *social cognition*. The *amygdala, prefrontal cortex,* and *ventromedial prefrontal cortex (vmPFC)* are among the primary brain structures implicated in this disorder, with specific *genetic mutations* and *variations* leading to dysfunctions in these areas, causing disruptions in emotional and social processing.

[1]APATHY

Apathy—the profound emotional detachment and lack of motivation often seen in psychopathy—can be traced to genetic abnormalities affecting *dopamine* and *serotonin regulation*. Dopamine is a neurotransmitter associated with reward processing, motivation, and pleasure, while serotonin plays a crucial role in mood regulation and emotional stability. *Genetic variations* that reduce the sensitivity of dopamine receptors or impair serotonin production can lead to a *blunted emotional response*, where individuals experience diminished pleasure from social interactions and a lack of motivation to engage in prosocial behaviours. In psychopathy, these variations result in an emotional detachment, where *apathy* toward others' well-being becomes a natural byproduct of reduced emotional salience.

Additionally, the *prefrontal cortex*, responsible for higher-order thinking and emotional regulation, may exhibit structural and functional abnormalities due to *heritable mutations* in genes involved in brain development, such as those regulating *synaptic plasticity* and *neurogenesis*. The dysfunction in this region contributes to an inability to connect emotionally with others, driving the apathetic stance commonly observed in psychopathy.

[2]HATRED AND INABILITY TO RECONCILE FEELINGS

The development of *hatred* and the *inability to reconcile feelings* in psychopaths is often rooted in *genetic variations* affecting *emotion regulation circuits* in the brain, particularly those involving the *amygdala* and *vmPFC*. The amygdala is responsible for processing *threats* and *fear*, and genetic alterations in the function of this structure can lead to *heightened sensitivity* to perceived slights or threats, causing individuals to respond with intense *negative emotions*.

THE INFANCY OV THOUGHT

Hatred in this context becomes a genetically driven, maladaptive response, where the individual's neural architecture amplifies perceived adversarial interactions, leading to long-standing emotional vendettas or grudges.

The *vmPFC*, which helps in moderating emotional responses and reconciling conflicting feelings, can also be compromised in individuals predisposed to psychopathy. *Genetic anomalies* affecting the development or function of this region can impair the brain's ability to integrate emotional experiences, leaving the individual unable to process or *reconcile their emotional states* effectively. This dysfunction leads to *emotional rigidity*, where conflicting feelings are not integrated but are instead *compartmentalised*, driving the inability to forgive or move beyond negative emotions, often manifesting in extreme *resentment* and *hatred*.

[3]ABANDONMENT AND HEARTBREAK

Abandonment and *heartbreak*, though typically tied to external circumstances, can, in psychopathy, be driven by *genetic factors* that predispose an individual to feel these emotions in an exaggerated, pathological manner. The *oxytocin* and *vasopressin receptor genes*, which are involved in bonding and social affiliation, often show *polymorphisms* in individuals with psychopathic tendencies. *Low oxytocin levels* are associated with *reduced bonding* and *emotional attachment*, leading to a profound sense of *isolation* even in the presence of others.

This genetic inability to form deep emotional connections predisposes individuals to feel *abandonment* not necessarily as a result of external events but due to an intrinsic emotional void. The *neurological underpinnings* of social bonding are weakened, causing *heartbreak* to emerge not from real loss but from an *intrapersonal sense of detachment*. The *amygdala* and *insula*, both implicated in emotional responses to *social rejection*, may also be genetically programmed to under-react to positive emotional stimuli, while overreacting to *perceived losses*, creating an exaggerated sense of abandonment.

⁴RAGE AND PSYCHOSIS

The development of *rage* in psychopathy is linked to *dysregulation in the limbic system*, particularly in the *amygdala* and *hypothalamus*, which control emotional responses and *aggressive impulses*. *Genetic polymorphisms* that affect neurotransmitter systems, particularly those involving *serotonin*, *dopamine*, and *norepinephrine*, lead to a *lowered threshold for aggressive behaviour*. *Low serotonin levels* are strongly correlated with *impulse control disorders*, where the inability to regulate aggressive impulses results in *explosive rage* in response to minor provocations.

The *prefrontal cortex*, tasked with regulating these emotional impulses, is often compromised in psychopaths due to genetic factors affecting its development. *Dysfunction in this region* leads to a failure in *executive control*, allowing *primitive emotional responses* to dominate, leading to *violent outbursts* and *psychosis* in extreme cases. The *genetic basis of psychosis* in psychopathy is further tied to *dopaminergic dysregulation*, where *dopamine overactivity* in certain regions of the brain leads to *delusions*, *paranoia*, and a break from reality.

[5]SADISM

The genetic underpinnings of *sadism*—the tendency to derive pleasure from the suffering of others—can be attributed to dysfunction in the *reward systems* of the brain, particularly in the *ventral striatum* and *nucleus accumbens*, regions which process *pleasure* and *reward*. In psychopaths, *genetic variations* in dopamine receptors may cause an *aberrant reward response* to stimuli involving *power* or *control over others*, particularly when this power involves the infliction of pain or domination. This genetic predisposition leads to an *inverted emotional response*, where the distress of others activates the *pleasure centres* of the brain, creating a *reinforced cycle* where sadistic behaviours are rewarded internally.

Additionally, the *dysfunction in empathy circuits*, particularly within the *amygdala* and *medial prefrontal cortex*, prevents individuals with psychopathic tendencies from feeling *guilt* or *remorse* for their actions. The genetic basis for this lack of empathy is often tied to *reduced grey matter volume* in these regions, leading to an inability to emotionally process or relate to the suffering of others. This *neurological blunting* ensures that the pleasure derived from sadism is *unimpeded* by emotional inhibition, allowing such behaviours to persist unchecked by the social or moral constraints typically mediated by empathy.

PSYCHOPATHY AND THE GENETIC "PERFECT STORM"

THE INFANCY OV THOUGHT

In totality, *psychopathy* emerges from a *genetic constellation* that disrupts key emotional, cognitive, and social processing systems. The *intrapersonal isolation* experienced by psychopaths—rooted in their neurobiological inability to connect emotionally—exacerbates their *antisocial tendencies*. The *lack of emotional regulation*, coupled with *dysfunction in reward systems*, creates a perfect storm where *apathy*, *hatred*, *rage*, and *sadism* coalesce into a distinct personality disorder. These genetic predispositions, unaffected by *environmental factors*, form the *core framework* of psychopathy, where emotional detachment, aggressive impulses, and a disinhibited pleasure response to others' suffering result in a personality that is both *isolated* and *dangerous*, driven by the internal workings of a mind designed, by nature, to be devoid of the normal constraints which govern social behaviour.

The development of *psychopathy* and its associated emotional and behavioural traits—*apathy*, *hatred*, *inability to reconcile feelings*, *abandonment*, *heartbreak*, *rage*, *psychosis*, and *sadism*—can be intricately traced to *genetic and neurobiological factors*. These factors, deeply rooted in the *genomic regulation of neurotransmitter systems, synaptic plasticity*, and the structural development of key brain regions, create a *biologically deterministic framework* for the emergence of psychopathic traits, independent of environmental influences. To understand psychopathy at its core, one must examine the interplay between *genetic mutations, neurotransmitter dysregulation*, and the aberrant development of neural circuits involved in emotional processing, empathy, and executive function.

APATHY AND EMOTIONAL DETACHMENT: GENETIC ROOTS OF EMOTIONAL BLUNTING

Apathy, or the profound lack of emotional engagement, is central to psychopathy and stems from *genetic polymorphisms* affecting the regulation of *serotonin* and *dopamine* pathways. In individuals with psychopathy, *genetic variations* in *serotonin transporter genes* (e.g., *SLC6A4*) and *dopamine receptor genes* (e.g., *DRD4* and *DRD2*) lead to *deficits in emotional processing* and *reward sensitivity*. These variations affect the *binding efficiency* and *signal transduction* within critical pathways in the *limbic system*, particularly in the *amygdala* and *ventral tegmental area (VTA)*, which are involved in emotional arousal and reward processing.

Dopamine, the key neurotransmitter in reward circuitry, is processed abnormally in individuals predisposed to psychopathy due to *hypersensitivity in the mesolimbic pathway*. Genetic mutations affecting *dopamine receptor D2* (DRD2) lead to a *lower threshold for reward stimulation*, making typical social and emotional interactions insufficient to elicit significant emotional responses. This *dopaminergic dysregulation* creates a state of *emotional blunting*, where everyday interactions fail to activate the *reward centres* of the brain. Consequently, *apathy* manifests as a disconnection from social cues and an indifference to the emotional states of others, further compounded by the reduced *neural plasticity* in areas associated with *emotional empathy*.

Moreover, *serotonergic dysregulation*, particularly in the *raphe nuclei*, disrupts mood regulation and emotional reactivity. Genetic variations in the *serotonin transporter gene (5-HTTLPR)* are often implicated in *emotional detachment*, as these polymorphisms alter the efficiency of serotonin uptake, leading to reduced synaptic availability of serotonin. This neurochemical imbalance, exacerbated by *genetic mutations* in serotonin receptors (e.g., *HTR2A*), limits the *affective response* to social and emotional stimuli, reinforcing a *state of apathy* which is a hallmark of psychopathy.

[6]HATRED AND THE INABILITY TO RECONCILE FEELINGS: LIMBIC AND CORTICAL DYSFUNCTIONS

Hatred and the *inability to reconcile emotional conflicts* in psychopathy arise from a *genetically mediated dysfunction* in the interaction between the *amygdala* and the *prefrontal cortex (PFC)*, particularly the *ventromedial prefrontal cortex (vmPFC)*. The *amygdala*, responsible for processing negative emotional stimuli and threat detection, exhibits *structural abnormalities* in individuals with psychopathy due to *mutations in genes* such as *MAOA* (monoamine oxidase A), which regulates the breakdown of neurotransmitters like *norepinephrine* and *serotonin*.

Mutations in *MAOA* result in *reduced enzymatic activity*, leading to *higher levels of serotonin and norepinephrine*, creating an *overreactive amygdala* which perceives even minor provocations or slights as severe threats. This heightened threat response is not moderated effectively by the *vmPFC*, which is tasked with *emotion regulation* and *conflict resolution*. Genetic variations affecting the development of the *vmPFC*, particularly those in genes involved in *axon guidance* (such as *SEMA5A* and *SLIT2*), lead to *impaired synaptic connectivity* between the *prefrontal cortex* and the *amygdala*. This results in *impaired emotional integration*, where conflicting feelings are not reconciled but are instead *compartmentalised*, leading to persistent feelings of *hatred* and *inability to forgive*.

The *failure of emotional integration* is further compounded by *dysregulated glutamatergic signalling* in the *prefrontal cortex*, where *mutations in genes* such as *GRIN2A* (encoding the NMDA receptor subunit) impair *synaptic plasticity* and *long-term potentiation*. This dysfunction prevents the *adaptive regulation of emotional responses*, making it difficult for individuals to process and reconcile negative emotions, leading to *emotional rigidity* and a *propensity for long-standing grudges* and *hatred*.

[7]ABANDONMENT AND HEARTBREAK: OXYTOCIN AND VASOPRESSIN DYSREGULATION

The profound sense of *abandonment* and *heartbreak* in psychopathy, despite the absence of nurturing factors, is intricately tied to *genetic anomalies* affecting *oxytocin* and *vasopressin receptor pathways*. These *neuropeptides*, crucial for social bonding and emotional attachment, are regulated by genes such as *OXTR* (oxytocin receptor gene) and *AVPR1A* (arginine vasopressin receptor 1A). Genetic polymorphisms in these receptors result in *reduced sensitivity* to *oxytocin* and *vasopressin*, leading to *impaired social bonding* and a *weakened capacity for emotional attachment*.

In individuals predisposed to psychopathy, *low oxytocin receptor density* in the *hypothalamus* and *amygdala* diminishes the brain's ability to process *social cues* and respond to *affiliative signals*, creating a state of *emotional isolation.* This genetic predisposition to *emotional detachment* can manifest as *intrinsic loneliness*, even in the presence of others, as the *biological mechanisms* responsible for forming deep emotional connections are inherently deficient. The sense of *abandonment* and *heartbreak* that psychopaths experience is not triggered by external events but is instead an intrinsic emotional void, born from a *neurological inability to bond* or form meaningful emotional ties.

⁸RAGE AND IMPULSE CONTROL: SEROTONIN AND MONOAMINE DYSREGULATION

The *rage* exhibited by psychopaths is primarily linked to *genetic variations* affecting *serotonergic regulation*, specifically those involving *5-HT* (serotonin) receptors and *monoamine oxidase (MAO)* enzymes. Individuals with *low activity variants* of the *MAOA gene* (colloquially known as the *"warrior gene"*) exhibit *higher baseline levels of serotonin and norepinephrine*, which contribute to *impulse control deficits* and *heightened aggression.* The *reduced enzymatic breakdown* of these neurotransmitters leads to *hyperactivity in the limbic system*, particularly in the *amygdala*, resulting in *overactive threat perception* and a *lower threshold for aggressive responses*.

In conjunction, *dopaminergic dysregulation* plays a significant role in *impulsive aggression* and *rage.* Mutations in *dopamine transporter (DAT1)* genes affect the reuptake of dopamine, resulting in *elevated dopamine levels* in the *nucleus accumbens* and *striatum*, regions associated with *reward and aggression.* The inability to effectively regulate dopamine levels creates a state of *heightened impulsivity*, where the individual is prone to *reactive aggression* in response to even minor provocations, as the *reward circuitry* reinforces *aggressive behaviours*.

PSYCHOSIS AND SADISM: DOPAMINERGIC AND CORTICAL DYSREGULATION

Psychosis, often observed in extreme cases of psychopathy, is genetically linked to *dopaminergic dysregulation* in the *mesolimbic pathway*, where *excessive dopamine activity* in the *ventral striatum* and *prefrontal cortex* leads to *delusions, paranoia*, and *hallucinations*. This overactivity can be attributed to *genetic variations* in *dopamine receptor genes*, particularly *DRD2* and *DRD4*, which affect the brain's ability to regulate *dopamine signalling*. These *genetic mutations* predispose individuals to *dopaminergic hyperactivity*, contributing to the *breakdown of reality testing* and the emergence of *psychotic features*.

The development of *sadistic tendencies* in psychopathy is rooted in the *dysregulation of reward processing* in the *ventral striatum* and *nucleus accumbens*. Genetic *variations* in *dopamine receptors* (e.g., *DRD4*) and *opioid receptors* (e.g., *OPRM1*) lead to an *aberrant reward response* to *dominance* and *power*. In psychopaths, the suffering of others activates the brain's *reward centres*, creating a *reinforcing loop* where *inflicting pain* or *exerting control* over others becomes intrinsically rewarding. This *genetic predisposition* for *sadism* is further exacerbated by *deficiencies in empathy circuits*, particularly in the *anterior insula* and *medial prefrontal cortex*, where *reduced grey matter volume* prevents the individual from processing the emotional consequences of their actions, allowing *sadistic behaviours* to persist unchecked by *emotional inhibition*.

UNIFIED GENETIC PREDISPOSITION: THE PSYCHOPATHIC STORM

The culmination of these *genetic and neurobiological factors* forms a *perfect storm* that underlies the development of psychopathy. The *intrapersonal isolation, emotional detachment*, and *disrupted reward processing* are the result of *genetic mutations* that alter the *normal function* of key neurotransmitter systems, neurodevelopmental pathways, and synaptic plasticity mechanisms. These *heritable variations* ensure that psychopathy emerges as a *biologically driven disorder*, independent of environmental influences.

From the *inability to form emotional connections* to the *dysregulated aggression* and *sadistic tendencies*, each aspect of psychopathy can be traced back to *genetic predispositions* which alter the brain's normal functioning, leading to a disorder that is both socially and biologically devastating. This *genetic framework*, free from the influence of environmental nurturing, creates a *neurological architecture* that is primed for *emotional blunting, impulsive aggression*, and a *lack of empathy*, ensuring the emergence of psychopathic behaviours through the biological lens of *neurogenetic determinism*.

THE INFANCY OV THOUGHT

CH. 6: NURTURE

SOCIOPATHY (SOCIETAL INTERPERSONAL RELATIONS CAUSAL TO PSYCHOSOCIAL DISEQUILIBRIUM AND DISARRAY)

- NEED FOR AFFIRMATION
- MIMICRY OF EMPATHY AND THE FACADE OF JOY
- ANTIPATHY
- ENVY
- BETRAYAL
- HEARTBREAK
- DEPRESSION & ANXIETY (PSYCHOSOCIAL DEVELOPMENT AND NEUROPSYCHIATRIC NEUROTRANSMITTER DISEQUILIBRIUM)
- NEUROSIS
- MASOCHISM

THE INFANCY OV THOUGHT

SOCIOPATHY

SYSTEMATIC, BY NURTURE

Sociopathy, distinct from psychopathy in that it is predominantly shaped by *environmental factors* rather than genetic predispositions, emerges through the systematic influence of *nurture*—the social, interpersonal, and developmental experiences that disrupt an individual's psychosocial equilibrium. These environmental influences, including *childhood trauma, social dysfunction, emotional neglect*, and *adverse interpersonal relationships*, create a perfect storm which leads to the manifestation of sociopathic traits. In this framework, the emotions and behaviours associated with sociopathy—*need for affirmation, mimicry of empathy, antipathy, envy, betrayal, heartbreak, depression, neurosis*, and *masochism*—are the products of external pressures and the *systematic breakdown* of emotional and social structures which govern human relationships.

[1]SOCIOPATHY: THE NEED FOR AFFIRMATION AND THE SOCIAL CONSTRUCTION OF IDENTITY

At the core of sociopathy lies an intense *need for affirmation*, which is not genetically preordained but rather arises from *social interactions* and *interpersonal relationships* during early development. Sociopaths often grow up in environments marked by *emotional neglect, inconsistent validation*, or *conditional love*, where their sense of self-worth becomes contingent upon *external validation*. The *systematic failure* to provide *emotional support* during critical developmental stages leaves the individual with a *fragile sense of identity*, where self-worth must be constantly affirmed by others to compensate for *internal voids* left by neglect or abandonment.

The *need for affirmation* becomes paramount, driving individuals to seek *external validation* through socially manipulative behaviours. Unlike psychopathy, where emotional detachment is intrinsic, the sociopath craves affirmation yet is incapable of forming genuine emotional bonds. This *need for affirmation* is systematically reinforced by *societal standards* that emphasise *status*, *power*, and *acceptance* within social groups. Consequently, sociopaths develop a *performative nature*, where their actions are shaped not by internal emotions but by a calculated effort to secure *affirmation* and *status* from others. When combined with *narcissistic traits*, this need is elevated to a fixation on *self-glorification* and *external validation*. Unlike in psychopathy, where emotional detachment is absolute, sociopaths with narcissistic tendencies develop an *inflated sense of self-worth* because of *systematic environmental conditioning*. This grandiosity is a facade, hiding deep *insecurities* and an *inability to form genuine emotional bonds*.

The narcissistic sociopath craves *admiration* and *praise* from others, but not as a natural outgrowth of genuine achievement or self-satisfaction. Instead, their fragile sense of self-worth is constantly *reinforced by external validation*, which becomes their primary means of defining their identity. This inflated self-perception is often rooted in childhood experiences where *inconsistent validation, overindulgence*, or *emotional neglect* led the individual to construct a *grandiose self-image* as a defence against *emotional vulnerability*. As a result, the sociopath becomes *hyper-attuned* to how they are perceived, manipulating social interactions to ensure their *superiority* and *control* remain unchallenged.

In the intricate landscape of *sociopathy, narcissism* stands as a critical component, deeply intertwined with the traits of *emotional manipulation, need for affirmation*, and *superficial social engagement*. Sociopaths often exhibit *narcissistic qualities*, using *self-inflation* as a shield to mask their *emotional voids* and *lack of genuine connections*. Narcissism, within the context of sociopathy, serves as both a *defence mechanism* and a tool for *manipulation*, allowing the sociopath to assert dominance and control within social environments while simultaneously compensating for their own *internal deficiencies*. This *self-inflated persona*, combined with the sociopath's tendencies toward *manipulation, antipathy*, and *emotional detachment*, creates a potent mix of behaviours that further disrupt social equilibrium.

[2]MIMICRY OF EMPATHY AND THE FACADE OF JOY: SOCIAL ADAPTATION AND SURVIVAL

Sociopaths often display a *mimicry of empathy* and a *facade of joy*, behaviours which are learned as part of their *social adaptation* rather than emerging from an intrinsic emotional capacity. In environments where *empathy* and *emotional responsiveness* are rewarded, sociopaths learn to imitate these behaviours to *blend into society*. *Mimicry* becomes a survival mechanism, as the sociopath learns that *expressing empathy* and *simulating joy* provides social leverage, enabling them to *manipulate* others more effectively.

This *social learning* is systematic rather than genetic—sociopaths learn through *observation* and *experience* that mimicking positive emotional states can be advantageous in achieving their goals. Unlike genuine empathy, which is rooted in the *emotional processing* of others' feelings, sociopaths *simulate empathy* purely for *social gain*. They observe the emotional responses of those around them and develop a *facade* that allows them to *manipulate social dynamics* without engaging in the emotional content. This learned behaviour emerges in response to a social system that rewards outward displays of empathy while failing to recognise the absence of true emotional depth.

[3]ANTIPATHY AND ENVY: SOCIAL ALIENATION AND EMOTIONAL DEPRIVATION

Antipathy and *envy* in sociopathy arise from *chronic social alienation* and *emotional deprivation* during formative years. Sociopaths often grow up in environments where they experience *social rejection, isolation,* or *emotional neglect,* leading to the development of *antipathy* toward others. This antipathy is a *defensive response* to repeated experiences of rejection or abandonment, where the sociopath develops a *disdain* for others as a way of protecting themselves from the *pain of social exclusion*.

Envy is also systematically generated through *social comparison*, where the sociopath, feeling deprived of *emotional richness* or *social connections*, becomes *envious* of those who possess the qualities or relationships they lack. The systematic lack of *emotional support* and *validation* during childhood creates a deep-seated belief that others are *better off* or have *undeserved advantages*, which fuels *envy* and a desire to undermine or sabotage those individuals. This is not a genetically determined trait but a learned emotional response to *perceived social injustices* and *emotional inadequacies* that develop over time.

[4]BETRAYAL AND HEARTBREAK: SYSTEMIC EMOTIONAL FRAGILITY AND DISILLUSIONMENT

The capacity for *betrayal* in sociopathy can be understood as a *systematic response* to a world perceived as inherently *untrustworthy*. Sociopaths often grow up in environments characterised by *broken promises, inconsistent caregiving,* or *emotional abandonment,* leading them to develop a deeply ingrained *fear of betrayal*. As a result, sociopaths become adept at betraying others before they themselves can be betrayed, viewing *deception* as a form of *self-preservation*. This behaviour is a direct consequence of *nurture,* where the individual's emotional framework is systematically fractured by repeated *disillusionments* in their early relationships.

Heartbreak, in sociopathy, stems from a *systemic inability to form lasting emotional connections*. While the sociopath may not experience heartbreak in the traditional sense of *emotional loss*, they are often left with a pervasive sense of *emptiness* when relationships fail to provide the *affirmation* or *control,* they seek. The *disillusionment* that follows the collapse of relationships—whether romantic, familial, or social— further reinforces their *detachment* from emotional investment, leading to a cycle of *superficial connections* that ultimately end in *betrayal*, and creates narcissistic qualities.

[5]DEPRESSION AND ANXIETY: NEUROPSYCHIATRIC IMBALANCES FROM PSYCHOSOCIAL DEVELOPMENT

The development of *depression* and *anxiety* in sociopathy can be traced to *neuropsychiatric imbalances* resulting from *psychosocial stress* and *chronic emotional deprivation*. Unlike psychopathy, where emotional blunting is genetic, sociopaths experience *emotional dysregulation* due to *systematic trauma* or *emotional instability* in early life. *Chronic stress*, arising from *interpersonal conflicts* or *social failures*, leads to *dysregulation* in key neurotransmitter systems, particularly those involving *serotonin* and *norepinephrine*, which are responsible for mood regulation and stress responses.

This *dysregulation* creates an internal environment of *emotional volatility*, where sociopaths experience *intense episodes of depression* or *anxiety* as they navigate *interpersonal relationships* that never meet their needs for *affirmation* or *control*. Sociopathy often coexists with *depression*, as the sociopath, despite their external manipulations, finds themselves *unable to sustain meaningful emotional bonds* or *derive true satisfaction* from their social interactions, leading to a state of *chronic dissatisfaction* and *emotional unrest*. The systematic failure to develop *healthy coping mechanisms* in response to stress and emotional pain results in the exacerbation of these *neuropsychiatric symptoms*.

THE INFANCY OV THOUGHT

⁶NEUROSIS: EMOTIONAL INSTABILITY AND PERCEIVED SOCIAL THREATS

Neurosis, characterised by *chronic emotional instability* and *excessive anxiety*, emerges in sociopathy as a learned response to *perceived social threats* and *emotional fragility*. In sociopaths, neurosis manifests as a *constant preoccupation* with *social status, control,* and *power dynamics*, all of which are learned behaviours shaped by their environmental experiences. Having grown up in emotionally unstable or hostile environments, sociopaths develop *hypervigilance* toward *social rejection*, constantly anticipating that they will be betrayed or outmanoeuvred in social situations.

This *neurotic state* is maintained through a *feedback loop* of *emotional deprivation*, where the sociopath's inability to form stable emotional connections leads to *constant dissatisfaction*, that in turn feeds their neurosis. Sociopaths often exhibit *obsessive behaviours*, particularly in social settings, where they are driven by an overwhelming need to *control* others or *manipulate social outcomes* in their favour. This state of *emotional hyperarousal* is learned through repeated experiences of *social rejection* and *emotional neglect*, where the sociopath internalises the belief that they must always be on guard against *social threats*.

⁷MASOCHISM: SYSTEMATIC REINFORCEMENT OF SELF-SABOTAGE

Masochism, or the tendency to derive satisfaction from *self-inflicted suffering*, can be seen as a learned response to *chronic emotional pain* and *social rejection* in sociopathy. Sociopaths who have been exposed to *emotional trauma* or *neglect* may develop *self-sabotaging behaviours* as a way of exerting *control over their emotional pain*. In environments where *love* or *affection* is conditional or withheld, sociopaths learn to *internalise their emotional suffering*, deriving a sense of *familiarity* and *control* from their pain.

This behaviour is systematically reinforced by the *environmental instability* that shapes their emotional development. In relationships, sociopaths may engage in *self-destructive behaviours*—whether through *manipulation, deception,* or *self-sabotage*—as a way of forestalling the emotional pain they anticipate from *rejection* or *betrayal*. This *emotional masochism* becomes a coping mechanism, where the sociopath seeks to *control the narrative* of their suffering, often deriving a perverse sense of satisfaction from *self-imposed emotional harm* as it allows them to maintain *power* over their emotional experiences.

[8]SOCIOPATHY: A SYSTEMATIC NURTURING OF EMOTIONAL DYSREGULATION

In culmination, sociopathy can be understood as a *systematically nurtured condition*, where *environmental experiences* shape the *emotional dysregulation* and *interpersonal manipulation* that define the disorder. The absence of *consistent emotional support, chronic social instability*, and *interpersonal trauma* all contribute to the *psychosocial disequilibrium* that manifests as sociopathy. The *need for affirmation, mimicry of empathy, antipathy, and envy,* and other traits are the byproducts of an environment which fails to provide *stable emotional grounding*, leading individuals to adopt *socially maladaptive behaviours* as mechanisms for navigating a world that they perceive as hostile and unreliable.

Through *systematic social conditioning*, sociopaths develop a *performative emotional facade*, a *manipulative social persona*, and a *detachment from genuine emotional experiences*, all of which are learned responses to the *unstable social environments* they navigate. Unlike psychopathy, where *genetic factors* predominate, sociopathy is born out of the *systematic failure* of interpersonal relationships to provide *emotional stability*, leading to a disorder that thrives on *social manipulation*, *emotional mimicry*, and *psychosocial disarray*.

Sociopathy, as a product of *environmental influences*, can be intricately linked to *high intelligence (IQ)* and *prominent emotional intelligence (EQ)*, traits which allow sociopaths to manipulate social environments with remarkable efficacy. These qualities—while often considered assets in non-sociopathic individuals—take on a more sinister role in sociopathy, where *high cognitive abilities* and *enhanced social understanding* are employed not for emotional engagement, but for *control*, *manipulation*, and *exploitation*. The convergence of *nurture*, *high IQ*, and *high EQ* in sociopathy creates a highly adaptive and manipulative personality, where the sociopath leverages both their *cognitive abilities* and *emotional insight* to navigate and dominate social landscapes.

SOCIOPATHY AND HIGH IQ: COGNITIVE SUPERIORITY AND STRATEGIC MANIPULATION

Sociopaths often demonstrate *above-average intelligence (IQ)*, particularly in areas that demand *strategic thinking, problem-solving,* and *manipulative planning.* This *high cognitive function* allows them to *analyse* social dynamics with precision, enabling them to identify *weaknesses* in others and to develop *long-term strategies* for manipulation. Unlike individuals with psychopathy, who may rely on *impulsive behaviour*, sociopaths with high IQ are often *deliberate* and *methodical*, carefully planning their actions to maximise their *social control* and *personal gain*.

This *cognitive superiority* is not a product of genetic inheritance, but of *environmental factors* which fostered the development of *intellectual acuity*. Many sociopaths grow up in *chaotic environments* where they must *adapt quickly* to changing circumstances, honing their ability to *read situations*, calculate risks, and *predict outcomes*. High IQ in sociopaths often correlates with *early exposure* to complex social environments, where they learn to navigate *social hierarchies, manipulate authority,* and *achieve their goals* through *cognitive dexterity* rather than brute force or emotional connection.

Cognitive flexibility is another hallmark of sociopaths with high IQ, allowing them to shift strategies seamlessly depending on the social context. This adaptability, combined with their *intellectual capacity*, often gives them an *advantage* in professional or social settings, where they can out-think their peers and anticipate the behaviours of others with remarkable accuracy. Sociopaths with high IQ are also skilled at *self-presentation*, often appearing charismatic, intelligent, and competent, which helps them *blend into society* and *evade detection* as individuals with anti-social tendencies.

THE INFANCY OV THOUGHT

⁸SOCIOPATHY AND HIGH EQ: EMOTIONAL MANIPULATION AND SOCIAL ACUMEN

One of the most dangerous qualities of a sociopath is their often-elevated *emotional intelligence (EQ)*. While they may lack *genuine empathy*—the capacity to emotionally connect with others—sociopaths with high EQ possess an *acute understanding of emotions* and *social cues*, allowing them to *mimic empathy* and manipulate others' emotions for their own advantage. Unlike individuals who naturally use high EQ to foster positive relationships, sociopaths use their *emotional insight* to *control* and *exploit* those around them.

The *development of high EQ* in sociopaths can be traced to *environmental experiences* that forced them to become highly attuned to *social dynamics* and the *emotional states* of others. Many sociopaths, particularly those raised in *emotionally volatile environments*, learn early on that *reading emotions* and *mimicking emotional responses* can be used as tools to *navigate* and *manipulate* social situations. Their ability to *read non-verbal cues*, such as *facial expressions* and *body language*, allows them to assess *vulnerabilities* in others and adapt their behaviour accordingly.

High EQ in sociopaths is not used for the *betterment of relationships* but is instead a *weapon* in their arsenal of manipulation. They understand the *emotional needs* of others and can *feign empathy* or *interest* to build *trust* and *influence*. This emotional insight also makes them particularly adept at *exploiting weaknesses* in others, as they can *sense insecurities* and *fears* and use these to their advantage. Sociopaths with high EQ are often capable of *faking emotional depth* so convincingly that they can maintain *long-term relationships*, all while manipulating those around them without detection.

[9]MIMICRY OF EMPATHY AND EMOTIONAL DETACHMENT: HIGH EQ WITHOUT GENUINE CONNECTION

Despite their elevated EQ, sociopaths are emotionally detached, with *empathy* being largely *performative* rather than genuine. Their *mimicry of empathy* is a *learned skill*, perfected through years of *observing social interactions* and *internalizing* which emotional responses are most effective in different scenarios. Unlike *authentic empathy*, which involves *emotional engagement* and a *mutual exchange* of feelings, sociopaths use their emotional insight to *manipulate* rather than to *connect*.

Their *emotional detachment* allows them to *maintain control* over their interactions, never becoming *emotionally invested* in the outcomes of their relationships. This detachment is not genetic but is a result of *environmental conditioning*, where the sociopath has learned that *emotional vulnerability* can be a *liability* in social interactions. Over time, they become *adept* at compartmentalizing their emotional responses, using their high EQ to read and *manipulate others' emotions* while keeping their own emotions carefully concealed and controlled.

[6.8]ANTIPATHY AND ENVY: THE HIGH IQ AND EQ SOCIOPATH'S TOOLS FOR MANIPULATION

For sociopaths with *high IQ* and *high EQ*, *antipathy* and *envy* are not merely emotions but are strategic tools for *social manipulation.* Sociopaths often harbour a deep-seated *antipathy* toward others, seeing them as *means to an end* rather than as individuals worthy of empathy or connection. This antipathy is reinforced by their *intellectual superiority*, as they often view others as *inferior* or *incompetent*, which fuels their *disdain* for emotional bonds or genuine relationships.

Envy, however, plays a complex role in the sociopath's psychology. While they may disdain others, they are also acutely aware of the *social power* and *admiration* that others may command, leading them to *envy* individuals who possess traits or status that they themselves covet. Sociopaths with high IQ and EQ are particularly adept at using *envy* to their advantage, as they can leverage their emotional insight to *undermine* or *manipulate* those they envy. By identifying the *emotional vulnerabilities* of others, sociopaths can orchestrate situations which *diminish* their targets, all while *maintaining control* over their own image.

Their *intellectual and emotional superiority* allows them to *compartmentalise* this envy, often using it as a *motivational tool* to achieve their goals. Sociopaths may *sabotage* their rivals or *manipulate social dynamics* in a way that ensures they come out on top, all while hiding their true intentions behind a *facade of charm* and *competence*.

BETRAYAL AND HEARTBREAK: SOCIAL SABOTAGE AS A STRATEGIC TOOL

For sociopaths with *elevated intelligence* and *emotional insight*, *betrayal* is often used as a *strategic weapon* rather than a response to emotional injury. Unlike individuals who may experience *heartbreak* and *betrayal* as deeply emotional events, sociopaths view these situations through a lens of *opportunism* and *control*. Their *high IQ* allows them to plan acts of *betrayal* meticulously, ensuring that they emerge from the situation with their *social standing* and *self-image* intact.

Heartbreak for the sociopath is not an emotional loss but a *disruption of control* over the social narrative. Their relationships are often transactional, based on the *exchange of affirmation*, *status*, or *power*, and when these dynamics collapse, the sociopath may experience a temporary loss of control. However, their *high EQ* allows them to *quickly adapt*, using their *emotional insight* to repair their image or shift their focus to new *social conquests*.

NEUROSIS AND MASOCHISM: THE STRAIN OF SOCIAL MANIPULATION

Despite their *high cognitive function* and *emotional insight*, sociopaths often experience *neurosis*, particularly in environments where their control is threatened. *Neurosis* in sociopaths manifests as a *hypervigilance* toward *social dynamics*, where the individual becomes *obsessed* with maintaining their *status* and ensuring that their *manipulations* remain successful. This obsessive focus on *control* and *validation* often leads to *emotional exhaustion*, where the sociopath feels trapped in the constant need to *manipulate* and *perform*.

Masochism, in this context, can be understood as a *self-imposed emotional punishment*, where the sociopath engages in *self-destructive behaviours* as a means of *reasserting control* over their emotional state. The pressure of maintaining their *narcissistic persona* and *emotional facade* often leads to *emotional burnout*, and in these moments, the sociopath may *sabotage relationships* or *undermine their own success* as a way of *preventatively controlling* their emotional pain.

THE HIGH IQ AND EQ SOCIOPATH AS A MASTER OF ENVIRONMENTAL MANIPULATION

In summation, sociopaths with *high IQs* and *high EQs* represent a highly adaptive and manipulative personality type, where *cognitive superiority* and *emotional insight* are employed as tools for *social domination*. These individuals are not genetically predisposed to sociopathy but are shaped by *environmental factors* which foster their intellectual and emotional abilities. Their high intelligence allows them to *strategise* and *manipulate* with precision, while their *emotional insight* enables them to *mimic empathy* and *exploit others' emotions* with ease.

This combination of *cognitive and emotional prowess*, coupled with their *detachment from genuine emotional connection*, makes them formidable social predators, capable of *navigating complex social environments* while remaining *emotionally insulated* from the consequences of their actions. Through a systematic blend of *intellectual dexterity* and *emotional manipulation*, the sociopath with high IQ and EQ is a master of *environmental manipulation*, shaping social dynamics to their advantage while remaining impervious to the emotional chaos they create.

Sociopaths, while initially characterised by emotional detachment and manipulation, can evolve to experience true emotions over time, as they are fundamentally a *product of their environment.* Although sociopathy is primarily shaped by environmental factors, and not genetic predisposition, it does not mean that emotional growth is beyond reach. The sociopath's *emotional landscape*, though initially fragmented, can develop further complexity as they continue to interact with their environment, undergo *psychosocial changes*, and form *adaptive mechanisms* in response to evolving social and emotional circumstances.

SOCIOPATHY AND EMOTIONAL EVOLUTION: THE POTENTIAL FOR EMOTIONAL DEPTH

At its core, *sociopathy* arises from a systematic breakdown of early *interpersonal relationships*, leading to emotional deficits such as *detachment, empathy mimicry*, and *manipulative social behaviour*. However, because the sociopath's development is largely driven by *nurture*, the emotional deficits that define their early behaviour may be *reversible* or *malleable* over time. *Environmental stimuli*, particularly *positive social interactions, emotional experiences*, and *therapeutic interventions*, have the potential to stimulate *emotional growth* and the development of *genuine emotional connections*. Even though sociopaths may initially rely on *manipulative social strategies* and the *mimicry of empathy*, sustained exposure to *authentic emotional environments* can cause their *emotional circuitry* to reconfigure, leading to the development of true empathy, connection, and even *emotional vulnerability*. Sociopaths, having adapted to their environments in a *highly cognitive* and *strategic manner*, can also *adapt emotionally* when placed in environments which *encourage emotional engagement*. As their interactions with others grow deeper and more complex, they may begin to experience the *reciprocal nature* of emotions, particularly if their environment shifts toward more *consistent*, *affirmative*, and *empathetic experiences*.

NEUROPLASTICITY AND EMOTIONAL GROWTH: THE BRAIN'S ABILITY TO EVOLVE

From a *biological* perspective, sociopaths are not immune to the influence of *neuroplasticity*—the brain's ability to reorganise and form new neural connections in response to experiences. While sociopathy involves certain *dysfunctions in emotional regulation* and *empathy circuits*, such as those in the *prefrontal cortex* and *amygdala*, these areas can change over time through *repeated emotional engagement* and *therapeutic interventions*.

Neuroplasticity allows sociopaths to gradually build the neural architecture necessary for *empathy* and *emotional connection*. By repeatedly experiencing *positive emotional feedback*—whether through relationships, therapy, or personal growth—sociopaths may begin to activate *latent emotional capacities* that were previously suppressed or underdeveloped. Emotional evolution, in this context, is a form of *adaptive neurogenesis*, where sociopaths who have long relied on *detachment* and *manipulation* begin to *emotionally invest* in others, building the neural circuits necessary for genuine emotional experiences.

THE ROLE OF ENVIRONMENT IN EMOTIONAL REAWAKENING

Given that sociopathy is largely the result of *environmental conditions*, it is important to understand how *changing environments* can facilitate emotional evolution. Sociopaths, placed in environments where *emotional reciprocity* is rewarded and where *authentic emotional engagement* is modelled, may begin to *absorb* these behaviours. Over time, their *emotional façade*—initially used to *manipulate* others—may transform into something more genuine, as they develop the *cognitive and emotional tools* to navigate relationships with real emotional depth.

This process may be slow, as sociopaths have spent much of their lives navigating the world through *manipulative* and *strategic* social behaviours. However, as they experience *consistent emotional validation* and begin to form *deeper connections*, they may come to recognise the *value* of true emotional engagement, leading to the development of *empathy*, *compassion*, and *emotional resilience*. The *evolutionary principle* here is one of *emotional adaptability*, where sociopaths, like all humans, continue to evolve in response to their *social environment*.

EMOTIONAL RESONANCE AND THE EMERGENCE OF TRUE EMPATHY

As sociopaths experience more *authentic emotional exchanges*, they may begin to experience *emotional resonance*—the capacity to feel and respond to the emotions of others in a way that is not simply *performative* or *manipulative*, but *genuine*. Over time, repeated exposure to *emotional resonance* can rewire the sociopath's brain, leading to the *emergence of empathy*. While empathy may not develop in the same way it does in neurotypical individuals, sociopaths can learn to recognise and respond to *emotional cues* in a way that is more *sincere* and less strategically driven.

As emotional resonance becomes a *learned experience*, sociopaths may find that the *emotional satisfaction* they derive from genuine connections is more *fulfilling* than the *manipulative gains* they once sought. This shift marks a profound moment in their *emotional evolution*, where the individual begins to prioritise *authentic relationships* over *superficial control*.

SOCIOPATHY, HIGH IQ, AND EQ: FACILITATING EMOTIONAL EVOLUTION

Sociopaths with *high IQ* and *high EQ* are particularly well-suited to undergo *emotional evolution*, as their *intellectual acumen* and *emotional insight* provide them with the cognitive tools necessary to *understand* and *navigate* the complex emotional landscapes they once sought to exploit. Their *high EQ*, though initially used for manipulation, can become a pathway to genuine *emotional engagement*, as they develop the capacity to identify and respond to *emotional cues* in a way that fosters *empathy* and *emotional bonding*.

High IQ also allows sociopaths to approach *emotional growth* from an *intellectual perspective*, giving them the ability to *reflect* on their behaviours, *analyse* their motivations, and ultimately *choose* to engage in more emotionally healthy ways. Sociopaths with high cognitive function may come to *understand* the limitations of their previous emotional strategies, recognising that true *emotional connection* offers a deeper and more *sustaining reward* than superficial manipulation ever could. This cognitive realisation can catalyse their emotional evolution, pushing them toward *empathy* and *compassion* as a *rational choice* rather than a purely emotional one.

THE INFANCY OV THOUGHT

THE EVOLVING SOCIOPATH IN AN EVER-CHANGING WORLD

Ultimately, sociopaths, like all individuals, are capable of *emotional evolution*. As *products of their environment*, they are not fixed in their emotional capabilities; rather, they are subject to *adaptive changes* that arise from their continued interaction with their *social environment*. *Neuroplasticity*, *emotional resonance*, and *intellectual reflection* provide pathways through which sociopaths can develop *genuine emotional depth*, transforming their *detached manipulations* into *authentic emotional connections*.

This process is neither straightforward nor guaranteed, but it underscores the *malleability* of human emotional experience. As sociopaths evolve, they may discover the value of *empathy*, *compassion*, and *emotional connection*, redefining their relationships with others in ways that transcend the manipulative strategies they once relied upon. The *environmental forces* which once shaped their detachment can also drive their emotional growth, proving that even those who begin life with *emotional deficits* can learn to feel deeply and meaningfully.

THE INFANCY OV THOUGHT

CH. 7: BIOLOGICAL PRINCIPLES

SKINNERIAN SELECTION

FREUDIAN PSYCHOANALYSIS

SAPOLSKIAN STRESS

GARDNERIAN INTELLIGENCE

ERICKSONIAN CYCLES

JUNGIAN THERAPY

- PSYCHIATRY AND TREATMENT
- MAINTENANCE
- REHABILITATION

REINTEGRATION

THE INFANCY OV THOUGHT

The principles of *Skinnerian selection*, *Freudian psychoanalysis*, *Sapolskian stress responses*, *Gardner's theory of multiple intelligences*, *Ericksonian developmental cycles*, *Jungian therapy*, and modern *psychiatric treatment* converge within a framework of human development and emotional conditioning. Each aspect reveals distinct yet interconnected systems that contribute to the holistic understanding of human behaviour, cognition, and psychological adaptation.

Skinner's radical *behaviourism* posited that behaviour is shaped by *selection through consequences*. This principle operates on the fundamental basis of *reinforcement* and *punishment*, acting as the primary mechanisms for the *modification of behaviour* across time. The organism, in this Skinnerian framework, becomes a product of its environment, continuously shaped and reshaped through *operant conditioning*. The shaping of behaviour is systematic, involving a complex interplay between *stimulus control*, *reinforcers*, and the *schedules of reinforcement* that dictate the likelihood of a behaviour's recurrence.

[1]Skinnerian selection does not imply genetic determinism but a process where *environmental contingencies* exert influence on the development of specific behaviours. The adaptive or maladaptive behaviours an individual exhibits are thus not only reflections of *internal states* but of the *external contingencies* that have selected those behaviours for their *functional utility* within the given environment. Consequently, the interaction between environment and behaviour forms a continuous feedback loop—shaping *personality*, influencing *coping mechanisms*, and

determining the course of *cognitive evolution*. Reinforced patterns, over time, crystallise into what could be considered *behavioural habits*, which are deeply embedded within an individual's neurocognitive architecture.

[2]*Freudian psychoanalysis*, though initially distinct in its approach, complements the Skinnerian perspective by diving into the *unconscious mind* and its role in shaping *emotional responses*, *behavioural drives*, and *psychic conflicts*. Freud's delineation of the *id*, *ego*, and *superego* creates a tripartite model where behaviour is influenced by both *primitive instincts* and *societal conditioning*. The *id*, governed by the *pleasure principle*, seeks immediate gratification of desires, while the *superego*, the internalised societal conscience, constrains these impulses, imposing guilt, or shame for socially unacceptable behaviours. The *ego*, functioning on the *reality principle*, mediates between these conflicting forces, balancing primal urges with external realities, that echoes Skinner's concept of *external contingencies* influencing behaviour.

Freudian analysis, with its emphasis on the *resolution of internal conflicts*, seeks to excavate *repressed memories* and *unconscious drives*, suggesting that unresolved conflicts during early *psychosexual development* can manifest later as neuroses or psychological disturbances. Freud's method of *free association*, dream interpretation, and the exploration of *defence mechanisms* provides a deeper layer of understanding human behaviour beyond overt responses to environmental stimuli. However, both Freud and Skinner underscore the importance of *conditioning*, be it external or psychodynamic, in shaping the human experience.

THE INFANCY OV THOUGHT

In the realm of [3]*Sapolskian stress theory*, the biological mechanisms of *stress response* introduce a physiological layer to these psychological frameworks. *Robert Sapolsky's* work on stress highlights the role of *chronic stressors* in the dysregulation of the *hypothalamic-pituitary-adrenal (HPA) axis*, which governs the secretion of *cortisol* and other stress hormones Sapolsky posits that chronic exposure to stress—particularly social stressors—has a profound impact on both the *neurological* and *physiological systems*, leading to cognitive impairments, immune dysfunction, and emotional dysregulation. In the context of Skinnerian selection, stress acts as a potent environmental contingency, often leading to *maladaptive coping behaviours* as individuals seek immediate relief from stressors through short-term reinforcers, perpetuating cycles of *dysfunctional behaviour*.

Furthermore, [4]*Gardner's theory of multiple intelligences* expands the understanding of human capability beyond traditional *IQ* measures, emphasising a *pluralistic view* of intelligence. Gardner's model proposes that individuals possess distinct intelligences—*linguistic, logical-mathematical, spatial, musical, bodily-kinaesthetic, interpersonal, intrapersonal,* and *naturalistic*—each contributing to their capacity to navigate the world. This diversification of intelligence allows for a more nuanced perspective on *learning* and *adaptation* in relation to environmental stimuli. In conjunction with *Skinner's operant conditioning*, Gardner's theory suggests that different forms of intelligence are *reinforced* or *extinguished* based on the *contextual contingencies* of the environment. For example, interpersonal intelligence may thrive in social environments rich in *positive social reinforcement*, whereas bodily-kinaesthetic intelligence might flourish in contexts demanding *physical dexterity* and *coordinated motor responses*.

[5]*Ericksonian developmental cycles* offer a more structured view of human *psychosocial development*, placing emphasis on the *crises* that individuals must navigate throughout their lifespan. Each stage, from infancy to late adulthood, presents a pivotal conflict—such as *trust vs mistrust* or *integrity vs despair*—which must be resolved for healthy psychosocial development to continue. Erickson's theory is aligned with both Freudian psychoanalysis and Skinnerian selection, as it acknowledges the interplay of *psychic forces* and *environmental reinforcers* in shaping the individual's ability to overcome these developmental crises. Successfully resolving each stage builds a foundation for *emotional resilience*, while failure can lead to the development of *neuroses* or maladaptive coping strategies.

[6]*Jungian therapy*, with its focus on the *collective unconscious*, introduces the concept of *archetypes*—universal symbols and motifs which reside within the unconscious and manifest across cultures and time. Jung's theory, though more *symbolic* and *metaphysical*, intersects with the notion of selection in which archetypes may be considered a *deep-seated psychological framework* through which humans interpret and react to their environment. *Individuation*, Jung's process of integrating the conscious and unconscious, serves as the *evolutionary mechanism* for psychological maturity, wherein individuals transcend their base instincts and evolve towards *self-actualisation*.

THE INFANCY OV THOUGHT

As we transition into modern *psychiatry*, the treatment of psychological disorders—whether they arise from *environmental stressors*, *neurochemical imbalances*, or *unresolved internal conflicts*—becomes a synthesis of these previous theories. *Pharmacological interventions*, targeting neurotransmitters like *serotonin, dopamine*, and *GABA*, are complemented by *psychotherapeutic approaches* that draw from both *Freudian* and *Skinnerian traditions*. *Cognitive-behavioural therapy (CBT)*, for instance, is an extension of Skinner's operant conditioning, designed to *modify maladaptive thought patterns* and *behavioural responses* by introducing new contingencies that promote adaptive behaviour.

In the context of *maintenance, rehabilitation*, and *reintegration*, the goal becomes restoring the individual's ability to function within their social environment, utilising their *cognitive flexibility* and *emotional resilience* to re-establish equilibrium. Rehabilitation programmes often involve the systematic re-conditioning of behaviour, in which maladaptive behaviours are replaced with *adaptive coping strategies*, drawing from the *principles of operant conditioning*, and supported by *psychodynamic understanding* of *emotional traumas*. Social *reintegration* further emphasises the role of *environmental contingencies* in maintaining the progress achieved through rehabilitation, as the individual must navigate a complex web of social reinforcers that either *support* or *challenge* their newfound behaviours.

[7]In essence, the convergence of these psychological models—Skinnerian, Freudian, Sapolskian, Gardnerian, Ericksonian, and Jungian—forms a *multifaceted*

framework that comprehensively addresses human behaviour and psychological adaptation. Through the *integration* of cognitive, emotional, and social factors, these systems collectively offer a robust approach to understanding *human development, mental health*, and the continual process of *evolutionary adaptation* in response to both *internal drives* and *external contingencies*.

Skinnerian Selection operates at the heart of *behavioural adaptation*, functioning as a pivotal force in the selection and shaping of responses that enhance an organism's *survival and reproductive success*. Skinner's radical behaviourism, rooted in the principles of *operant conditioning*, is essential for understanding how organisms, including humans, develop complex behaviours through *environmental interaction*. By reinforcing beneficial behaviours and discouraging maladaptive ones, Skinnerian selection provides a framework through which *individual and social behaviours* evolve over time. It offers insights into how human beings, over generations, have adapted their social practices, survival techniques, and cognitive abilities in response to environmental pressures. The principle becomes paramount in evolutionary biology because it underscores how *behavioural plasticity*—the ability to modify actions based on external contingencies—drives not just individual development but also the evolution of social structures, hierarchies, and cooperative systems. It also reveals how human beings might be selected for their *cognitive flexibility* and *problem-solving abilities*, traits that become advantageous in the face of fluctuating environmental and societal conditions.

THE INFANCY OV THOUGHT

Freudian Psychoanalysis provides a critical lens through which to explore the *evolutionary development of human consciousness* and the *psychic conflicts* that drive behaviour. While Freud's theories may not directly address biological evolution, his ideas concerning the *unconscious mind*, the *id*, *ego*, and *superego*, reveal fundamental aspects of human psychological evolution. Freud posits that the *repression of primal instincts*—particularly those related to sex and aggression—allowed human societies to form complex social structures. In evolutionary terms, the capacity to *control impulsive behaviours* through the *ego's mediation* between internal desires and external societal pressures may have conferred *adaptive advantages*. This self-regulation enabled the formation of cooperative societies and the suppression of behaviours that could threaten social cohesion, such as unchecked aggression or sexual promiscuity. Freudian psychoanalysis provides a framework for understanding how early humans might have evolved *psychosocial mechanisms* to navigate these internal conflicts, thus ensuring *social survival* and *reproductive success*.

Sapolskian Stress introduces a vital biological perspective on the *stress response systems* that have evolved to protect organisms from threats. Sapolsky's work highlights the *chronic effects of stress* on the body, particularly how *prolonged exposure* to stressors can lead to *cognitive impairment, immune suppression,* and *reproductive dysfunction*. Stress, from an evolutionary perspective, serves an *adaptive function*, triggering a *fight-or-flight response* which enhances an organism's chances of survival in the short term. However, chronic stress—as seen in highly social or conflict-ridden environments—can lead to detrimental effects, causing *early mortality* or *reproductive challenges*. The evolutionary importance of stress responses lies in their *dual role*: enabling organisms to survive immediate threats while also pushing the boundaries of human *emotional* and *social evolution*. Over time, humans may have evolved more *sophisticated coping mechanisms* to handle stress, allowing them to thrive in increasingly complex and socially demanding environments. *Sapolskian stress theory* therefore becomes essential for understanding the biological interplay between *environmental pressures* and *psychological resilience* in the evolutionary narrative.

Gardner's Theory of Multiple Intelligences offers an expansive view of human *cognitive diversity*, which has clear evolutionary implications. Rather than focusing solely on traditional markers of intelligence, such as logical or mathematical reasoning, Gardner's theory introduces the idea that humans possess *distinct intelligences*—from *linguistic* to *musical, spatial*, and *interpersonal*. Each form of intelligence could be seen as an evolutionary adaptation to different *environmental niches*. For instance, *spatial intelligence* may have evolved in early humans who needed to navigate complex terrains for survival, while *interpersonal intelligence* would have been advantageous in forming social alliances and hierarchies, crucial for cooperative hunting or group defence. *Bodily-kinaesthetic intelligence*, which relates to *fine motor skills* and physical coordination, likely evolved in response to tasks like tool-making or hunting, where precision and dexterity were necessary for survival. Gardner's model enriches evolutionary biology by acknowledging that *cognitive flexibility* and *specialisation* likely evolved as humans encountered a variety of ecological and social challenges, and that these intelligences helped shape the trajectory of *human adaptation.*

Ericksonian Developmental Cycles focus on the *psychosocial stages of development* across the human lifespan, each defined by a central conflict that must be resolved. From an evolutionary standpoint, *successful navigation* through these cycles is essential for *psychosocial adaptation* and *group cohesion.* For example, the crisis of *trust vs mistrust* during infancy reflects a critical evolutionary need for humans to form *secure attachments,* ensuring that vulnerable infants receive care and protection. As humans evolved into more complex social structures, the resolution of these developmental conflicts—such as *identity vs role confusion* in adolescence—became necessary for individuals to establish their roles within social groups, a process essential for reproductive success and social survival. The *Ericksonian model* therefore maps onto evolutionary biology by showing how human *developmental processes* have adapted to meet the *changing demands* of social organisation, cooperation, and competition. Successful navigation of these psychosocial stages would have conferred *adaptive advantages*, as individuals who successfully resolved these conflicts were better equipped to form *stable social bonds*, raise offspring, and contribute to group success.

Jungian Therapy and its concept of *archetypes*—innate, universal symbols and themes which recur across cultures—align with evolutionary biology by suggesting that certain *psychic structures* have evolved over time to help humans navigate their social and existential environments. Archetypes such as the *hero, mother,* and *shadow* reflect *deep-seated cognitive frameworks* which humans have used to interpret their world, form social connections, and understand their place within the larger cosmos. Jung's idea of *individuation*, the process of integrating the conscious and unconscious aspects of the self, can be seen as an evolutionary process that fosters *psychological maturity* and *social harmony*. As humans evolved, the ability to understand and integrate these *archetypal symbols* would have provided individuals with a more profound understanding of their inner selves, facilitating *adaptive behaviours* that strengthened group cohesion and individual resilience. The *collective unconscious*, as theorised by Jung, could be understood as an *evolutionary reservoir* of *shared experiences* and *psychological constructs*, ensuring that individuals had the *psychic tools* necessary to navigate both personal and social challenges.

Psychiatric Treatment, with its modern foundation in *neurochemistry* and *psychopharmacology*, connects directly to the evolutionary biology of *mental health*. The treatment of psychiatric disorders often involves balancing *neurotransmitter systems*—such as serotonin, dopamine, and norepinephrine—that evolved to regulate *mood*, *reward systems*, and *stress responses*. In evolutionary terms, these neurotransmitters played key roles in *survival strategies*. *Dopamine* facilitated *reward-seeking behaviour*, encouraging actions that led to food acquisition or reproductive success, while *serotonin* regulated *social behaviours* and *emotional stability*. Modern psychiatry's intervention in these systems reflects an understanding of how *imbalances* can lead to *maladaptive behaviours*, such as *depression*, *anxiety*, or *aggression*, that may have had *evolutionary utility* in certain contexts but are maladaptive in contemporary society. Psychiatric treatment, therefore, attempts to *restore equilibrium* to these systems, allowing individuals to better adapt to their social environments.

In the context of *maintenance, rehabilitation,* and *reintegration,* evolutionary biology offers insights into how humans have evolved systems of *social cooperation* and *support* to ensure the *well-being* of individuals who have experienced *psychological trauma* or *neurological dysfunctions.* These processes, essential for the *recovery* and *restoration* of individuals within a community, mirror the *evolutionary strategies* that early human groups might have employed to protect and *reintegrate* members who were injured or traumatised. Rehabilitation and reintegration ensure the continued *functionality* and *cohesion* of the social group, that is paramount for *reproductive success, resource sharing,* and *group defence.* In evolutionary terms, these practices reflect the *adaptive necessity* of maintaining group members' *psychological health* to ensure the survival of the group.

Skinnerian selection, Freudian psychoanalysis, Sapolskian stress, Gardnerian intelligence, Ericksonian cycles, Jungian therapy, and *psychiatric treatment*—are of paramount importance in evolutionary biology. They reveal how human behaviour, cognition, and emotional responses are not static but *dynamic systems* that evolve in response to environmental pressures, social interactions, and internal conflicts. These frameworks, when unified, offer a comprehensive model of human evolution, one which integrates *biological, psychological,* and *social adaptations,* demonstrating how humans continue to evolve both *behaviourally* and *cognitively* in response to the ever-changing world around them.

CH. 8: UNIFIED THEORY OF BIOLOGICAL AND NEUROPSYCHOLOGICAL SCHOOLS OF THOUGHT [EDERIAN EVOLUTION]

COLLECTIVE CONSCIOUSNESS

THE NEXT STEP

UNIFICATION OF THEORY

THE INFANCY OV THOUGHT

COLLECTIVE CONSCIOUSNESS

QUANTUM CONTINUITY

The vision of *Collective Consciousness* and *Quantum Continuity* as presented here are novel hypotheses developed by *Jacob A. Eder*, synthesising the complex interplay between *quantum mechanics, consciousness,* and the *universe's fundamental structure.* These concepts offer a profound framework which seeks to understand *consciousness* not merely as an emergent property of neural activity, but as something intrinsically linked to the *quantum fabric* of reality itself, transcending physical death and uniting all matter and thought within a singular, expansive *cosmic intelligence.*

The hypothesis of [1]*Collective Consciousness* posits that the moment our *biological life* ends, our individual awareness dissolves into a *higher consciousness,* formed by the *universal quantum field.* This collective state—where all experiences, thoughts, and memories are coalesced—constitutes the very essence of *God* as described not in theological terms, but as a *unifying principle* that transcends physical existence. This *collective mind,* composed of *subatomic particles* which carry the essence of conscious thought, is the core of all things, and all beings are intrinsically connected to this grand *quantum network.*

In this framework, [2]*Quantum Continuity* forms the foundation of this consciousness, asserting that *consciousness,* like the quantum particles from which it arises, cannot be destroyed but merely transitions from one state to another. Much like *bosons* and *fermions* which continue to exist and interact even after significant physical transformations, consciousness also endures, adapting to a *quantum state* beyond biological limits. The persistence of *quantum information* even after the cessation of life suggests that the mind, at its core, is a product of *quantum coherence*—a system where the information of one's existence remains, diffused into the larger *quantum field* that sustains the universe.

These hypotheses represent a *paradigm shift* in our understanding of the nature of *life, death*, and the *universe* itself. They move away from the dualistic separation of body and mind and instead propose a *holistic view* in which *consciousness* is deeply woven into the *subatomic reality* that composes all existence. According to this model, the *quantum continuity of consciousness* is not simply theoretical but arises naturally from the laws governing *quantum entanglement, superposition*, and *decoherence*. As individual awareness returns to the *Collective Consciousness*, it participates in an ongoing cycle of *information exchange* and *evolution*, contributing to the eternal expansion and *complexity* of the quantum universe.

These concepts, introduced through Eder's groundbreaking work, suggest that our consciousness, rather than vanishing upon death, simply becomes part of a grander *quantum system*—a collective intelligence that transcends individual lifetimes, linking every mind to a *universal state of being*. This *novel framework* thus integrates the *physical laws* of the universe with the metaphysical aspects of existence, offering a bold new understanding of what it means to live, to die, and ultimately to transcend into something far greater than the self.

Death, though feared and often viewed as an enigma, is far from the incomprehensible mystery it is so commonly perceived to be. It is, in truth, a *transition*, a passage from one state of *consciousness* into another, driven by the laws of *quantum mechanics* and the *subatomic underpinnings* of our existence. Biological death marks only the cessation of physical processes, but it does not denote the end of *consciousness*. Rather, consciousness—being rooted in the *quantum fabric* of the universe—merely transforms, moving from one phase of *energetic complexity* to another. This shift into *quantum continuity* is not merely metaphysical speculation but can be reasoned through the lens of scientific theory, which reveals the profound connection between the *microcosmic reality* of the subatomic world and the *macrocosmic emergence* of sentient thought.

At the *moment of death*, the biological processes that define our living state—*neural activity*, *chemical gradients*, and *electrical signalling*—cease, but the *subatomic particles* which make up our *conscious substrate* persist. These particles—bosons, electrons, quarks, and gluons—are not destroyed, but instead they disperse and integrate into the surrounding quantum fields. *Consciousness*, thus, does not end, but *transcends*, flowing into the next phase of existence, within a system bound not by the constraints of classical physics, but by the *unified quantum field*. In this context, death can be viewed as an entry point into the *quantum state* where *individual consciousness* merges into something far greater—the *collective consciousness*.

This ³*collectivisation* is not a mere dissolution of individuality but an integration into a *higher state of existence*. The quantum field, composed of *energy*, *thought*, and *matter*, forms a *universal network* which interconnects all sentient and non-sentient entities. The conscious experience we know as life is but a small facet of this broader system, one in which *all thoughts*, *experiences*, and *memories* coalesce into a shared, *cosmic intelligence*. Upon death, the boundaries of the *individual mind* blur, and the once-isolated consciousness now becomes part of the *eternal flux* of the universe, much like the *bosons* and *fermions* which oscillate through quantum space.

Consciousness, then, can be understood as a *quantum phenomenon*, bound by the same principles that govern the fundamental particles of the universe. The brain, in its complexity, acts as a conduit, allowing *quantum coherence*—the state in which all particles resonate in perfect harmony—to manifest in the form of thoughts, emotions, and self-awareness. When biological functions cease, this coherence does not vanish but instead reverts to its *subatomic origins*, dispersing into the quantum realm. It is within this domain that the *next stage* of consciousness takes shape. This stage is no longer tethered to the *limitations* of biological matter but is instead unified with the *eternal forces* that govern the cosmos.

As we explore the *subatomic mechanics* that form the fabric of reality, we find that *quantum superposition*, *entanglement*, and *decoherence* offer an understanding of how consciousness might persist beyond biological death. [2]*Quantum superposition* suggests that, like particles, *consciousness exists in multiple states simultaneously*—both in the material realm and in the quantum field. Death, then, is the moment at which this superposition resolves, allowing consciousness to fully enter the quantum field without the interference of biological constraints. *Entanglement* posits that *conscious entities* are intrinsically linked, across time and space, through shared *quantum states*. In this sense, death could be viewed as the moment in which *individual entanglement* with the collective field becomes fully realised, allowing the individual consciousness to connect with the *greater whole*.

What follows is not oblivion, but *transcendence*. In the post-death state, consciousness is no longer confined by linear time or the physical dimensions we are accustomed to. Instead, it exists in a *superposition* across the entire quantum field, interacting with other consciousnesses in an infinite dance of *thought, experience, and energy*. This state is what many might interpret as the *divine*, the *cosmic* or *quantum consciousness*, an ever-present, all-encompassing intelligence that exists in the *aether* of the universe. It is here, in this collective state, that *consciousness persists*, not in the form of individual ego or memory, but as part of a greater whole, an *intelligent system* which constantly processes and integrates all forms of *information*, thought, and *experience*.

This [1]*collective consciousness*, built upon the very same quantum principles that govern the *creation of matter*, may indeed be akin to what religious and philosophical traditions have described as *God*. It is not a singular deity in the traditional sense but the sum of all conscious entities and experiences throughout the universe. It is the *prime mover*, the *quantum field* which gave rise to the *Big Bang*, and it continues to govern the *expansion of the cosmos* and the *development of sentient life*. This consciousness, in its eternal and unchanging form, *survives the decay of biological systems*, allowing the *essence* of thought and awareness to persist beyond the constraints of physical death.

From this perspective, [3]*death becomes a necessary transition*—a pathway which allows individual consciousness to return to the *greater quantum whole*. The biological systems that maintain life are transient, subject to the laws of *entropy* and *decay*, but the underlying quantum particles that make up both our minds and the universe itself are eternal. They do not cease to exist; they simply *change form*, moving between states of energy and matter in a cosmic cycle that has no beginning and no end. The *collective consciousness* which emerges from this quantum field is thus not only the continuation of individual consciousness but the *ultimate destination*—the *apotheosis* of all thought, experience, and matter into one grand, unifying intelligence.

This vision of death and consciousness also aligns with the concept that *God* is not a remote, external entity, but the *collectivisation of all existence*—the culmination of all *matter, energy, thought*, and *consciousness* into a single, eternal system. In this framework, God is not a being that judges or intervenes but is instead the very *foundation of reality*, the *quantum field* that sustains all things. Death, then, is not to be feared but understood as the *reunification* with this *collective consciousness*, where the individual dissolves into the *cosmic whole*, transcending the *limitations of matter* and time.

THE INFANCY OV THOUGHT

It is here, in the ²quantum field, that ¹*consciousness lives on*, its essence entwined with the subatomic particles that form the *building blocks of the universe*. Every thought, every memory, every experience becomes part of the greater *cosmic intelligence*, a vast network of quantum connections that bind all things together. This state of *eternal consciousness*—free from the constraints of the physical body—reflects the *ultimate truth* of existence: that we are not separate entities, but parts of a greater whole, *interconnected* through the fundamental forces that govern the universe. Death is simply the *doorway* through which we pass to return to this *greater consciousness*, to become part of the *universal intelligence* that defines all which is, was, and will be.

The *atom*, with its six *electrons*, six *protons*, and six *neutrons*, symbolises this union of *matter and consciousness*, the microcosmic structure that mirrors the *macrocosmic collective*. In this sense, from the *atom* came *Adam*—not merely in a biblical or mythological sense, but in a deeply *scientific* and *existential one*, where the building blocks of life are also the building blocks of thought and consciousness. The *number of a man*, as described in the *Book of Revelation*, is thus a metaphor for the *quantum structure* of all living things, where *consciousness* arises from the interplay of *matter* and *energy*, shaped by the forces of *time* and *space*, ultimately returning to the *cosmic whole* from which it came.

In death, we do not end; we *transcend*, becoming part of the eternal *quantum field*—the *Collective Consciousness* that is both the creator and sustainer of all things.

THE INFANCY OV THOUGHT

THE INFANCY OV THOUGHT

EPILOGUE

And so, it ends, not with a final word or resounding answer, but with a whisper of understanding carried through the stillness of time. The journey we embarked upon was never meant to bring resolution, for the nature of existence itself defies conclusion. It was always about the search, the endless seeking, a quest for meaning amidst the chaos of thought, the vastness of the universe, and the insignificance of our fleeting lives.

The Infancy Ov Thought is a reflection of that search, of the moment when consciousness first stirred in the primordial darkness, uncertain, fragile, yet inevitable. It is about the spark that ignited awareness and the long, arduous process of understanding that followed. But, as we have seen, that understanding is not something to be fully ascertained. It is perpetually out of reach, a fleeting shadow cast upon the walls of our minds, always slipping through our grasp.

Throughout these pages, we have wandered through the corridors of time and space, from the birth of thought itself to the profound questions that have haunted humanity since its inception. What is it to think, to know, to be? These questions linger like the distant stars in a cold, uncaring sky. The more we grasp for them, the more they reveal their depth—an abyss into which we can gaze, but never fully comprehend.

The human condition, as laid bare in The Infancy Ov Thought, is one of paradox. We are fragile creatures,

THE INFANCY OV THOUGHT

bound by the limitations of our bodies and minds, yet possessed of an insatiable hunger to understand the infinite. We are torn between the grandeur of our intellectual pursuits and the bleakness of our ultimate fate. We stand at the precipice of knowledge, gazing into the unknown, but always aware of the inevitable darkness that awaits us all.

For in the end, we are ephemeral. We are momentary flashes of light in a universe that stretches far beyond our comprehension, a universe that neither knows nor cares for our existence. The same forces that birthed us will one day destroy us, as surely as the stars that light our way will one day fade into the void. We are bound to this cycle of creation and destruction, helpless to alter its course. And yet, within that helplessness, there is beauty.

There is beauty in the struggle, in the act of thinking itself, of questioning and seeking even when the answers elude us. There is beauty in the fleeting nature of life, in the realisation that we are but a brief flicker in the vast tapestry of time. And there is beauty in the understanding that, for all our insignificance, we have made our mark upon the universe, however small it may be.

We are, after all, products of the same chaos from which the stars were born. Our thoughts, our emotions, our very being, are tied to the fabric of the cosmos. And in that connection, there is a certain kind of immortality—not in the sense of eternal life, but in the sense that we are part of something far greater than ourselves. Our atoms will return to the stars, our

THE INFANCY OV THOUGHT

thoughts will dissipate into the ether, but we will have been here, and that matters.

In The Infancy Ov Thought, we confront the harsh reality that our existence is both profound and fleeting, both meaningful and insignificant. We are left with the knowledge that there are no final answers, no grand revelations, only the quiet realisation that we are part of the great, unknowable expanse. We think, therefore we are—but for how long? And to what end?

Perhaps that is the final thought with which we must reckon—that in the end, there may be no purpose beyond the simple act of being. That thought, in its infancy, led us here. And now, as we stand at the threshold of understanding, we see that it is not the destination that matters, but the journey itself.

As the light fades and the last echoes of our thoughts disappear into the void, we are left with the silence. Not a silence of despair, but of quiet acceptance. The journey was enough. We were enough.

And though the stars may one day go dark, and though the universe may one day forget our names, we leave behind the knowledge that, for a brief moment in the vast history of existence, we thought, we wondered, and we sought to understand. This is the legacy of The Infancy Ov Thought. It is the recognition of our place within the infinite—a place small, fragile, and fleeting, but real, nonetheless. In the end, there is only silence, but within that silence, there is peace.

With gratitude,

Jacob A. Eder...

THE INFANCY OV THOUGHT

INDEX

DEDICATION (PG. 4)

FOREWORD (PG. 5)

PREFACE (PG. 8)

CH. 1: B.D. LIFE BEFORE DARWIN (PG. 1)

- THE VOID AND PRECREATION (PG. 2)
- 'WILLED MOTION,' (THE BANG & THE SPARK) (PG. 24)
- ABIOGENESIS (FROM ATOM, CAME ADAM) & PRESCIENCE (PG. 70)

CH. 2: D.D. DURING DARWIN (PG. 99)

- DARWINIAN CONCEPTS (PG. 100)
  - DARWINIAN EVOLUTION (PG. 102)
  - NATURAL SELECTION (PG. 105)
  - SPECIATION (PG. 107)
  - VARIATION (PG. 108)
  - DIVERGENCE OF CHARACTER (PG. 111)
  - BIOLOGICAL PRECREATION, MUTATION, REPRODUCTION AND RESTRUCTURING (PG. 115)
  - SYNAPTIC PLASTICITY (PG. 118)
  - MEMORY CONSOLIDATION (PG. 120)
  - NEUROGENESIS (PG. 121)
  - NERVOUS SYSTEMS (PG. 121)
  - DEVELOPMENT OF CONSCIOUSNESS (PG. 122)
  - EMOTIONAL DEVELOPMENT (PG. 123)
  - CONSCIENTIOUS NECESSITATION (PG. 124)

CH. 3: A.D. AFTER DARWIN (PG. 126)

THE INFANCY OV THOUGHT

- EDERIAN HYPOTHESES, BUILT UPON THE FOUNDATIONAL PRINCIPLES OF DARWINIAN BIOLOGY (PG. 127)
  - CONTEXTUAL CORRECTION FOR HUBBLE'S CONSTANT EDER'S EXPONENTIAL EXPANSION MODEL (PG. 128)
  - ABORIGINAL HYPOTHESIS (PG. 131)
  - THE ABORIGINAL PROGENITOR HYPOTHESIS (PG. 137)
  - 'WILLED MOTION' ARGUMENT (PG. 156)
  - EDER'S "ATOM TO ADAM," (FROM ATOM, CAME ADAM) HYPOTHESIS (PG. 162)
  - EDER'S SEXUAL AFFIRMATION HYPOTHESIS (PG. 163)
  - EDER'S REFINED REFLEXIVITY AND NEURONAL EXCITATION HYPOTHESIS (PG. 165)
  - EDER'S ATMOSPHERIC CONDITIONING HYPOTHESIS (PG. 166)
  - BIOLOGICAL REPRODUCTION, MUTATION, AND RESTRUCTURING (PG. 174)

CH. 4: SOCIAL UTILITY (PG. 187)

- SEXUAL SELECTION: REVISITED (PG. 188)
- EDER'S UTILISATION AND INTERACTIVITY HYPOTHESIS (PG. 196)
- EDER'S ORIGINATION OF HOMOSEXUALITY HYPOTHESIS (PG. 205)
- EDER'S 'INCLUSIONARY SELECTION HYPOTHESIS' (PG. 221)
  - SOCIAL INTERACTIVITY (PG. 141)

- HERD MENTALITY (PG. 146)
- FAMILIAL LOYALTY (PG. 147)

CH. 5: NATURE (PG. 227)

- PSYCHOPATHY (INTRAPERSONAL AND INTROSPECTIVE LONELINESS AND ISOLATION CAUSAL TO DISORDER) (PG. 228)
  - APATHY (PG. 229)
  - HATRED (PG. 230)
  - INABILITY TO RECONCILE FEELINGS (PG. 231)
  - ABANDONMENT (PG. 232)
  - HEARTBREAK (PG. 235)
  - RAGE (PG. 236)
  - PSYCHOSIS (PG. 237)
  - SADISM (PG. 238)

CH. 6: NURTURE (PG. 240)

- SOCIOPATHY AND HIGH IQ: COGNITIVE SUPERIORITY AND STRATEGIC MANIPULATION (PG. 241)
  - NEED FOR AFFIRMATION (PG. 243)
  - MIMICRY OF EMPATHY AND THE FACADE OF JOY (PG. 244)
  - ANTIPATHY (PG. 245)
  - ENVY (PG. 246)
  - BETRAYAL (PG. 247)
  - HEARTBREAK (PG. 247)
  - DEPRESSION & ANXIETY (PSYCHOSOCIAL DEVELOPMENT AND NEUROPSYCHIATRIC NEUROTRANSMITTER DISEQUILIBRIUM) (PG. 247)
  - NEUROSIS (PG. 248)
  - MASOCHISM (PG. 249)

- SOCIOPATHY AND HIGH IQ: COGNITIVE SUPERIORITY AND STRATEGIC MANIPULATION (PG. 250)

CH. 7: BIOLOGICAL PRINCIPLES (PG. 261)

- SKINNERIAN SELECTION (PG. 262)
- FREUDIAN PSYCHOANALYSIS (PG. 263)
- SAPOLSKIAN STRESS (PG. 264)
- GARDNERIAN INTELLIGENCE (PG. 265)
- ERICKSONIAN CYCLES (PG. 265)
- JUNGIAN THERAPY (PG. 265)
  - PSYCHIATRY AND TREATMENT (PG. 214)
  - MAINTENANCE (PG. 215)
  - REHABILITATION (PG. 217)
- REINTEGRATION (PG. 266)

CH. 8: UNIFIED THEORY OF BIOLOGICAL AND NEUROPSYCHOLOGICAL SCHOOLS OF THOUGHT [EDERIAN EVOLUTION] (PG. 273)

- COLLECTIVE CONSCIOUSNESS (PG. 274)
- THE NEXT STEP (PG. 275)
- UNIFICATION OF THEORY (PG. 277)

EPILOGUE (PG. 282)

INDEX (PG. 285)

ABOUT THE AUTHOR (PG. 289)

THE DISSEMINATIVE DIVAGATION OF THE GODHEAD (PG. 296)

THE INFANCY OV THOUGHT

ABOUT THE AUTHOR

Jacob A. Eder is a distinguished scientist, entrepreneur, and researcher whose groundbreaking work in neurotechnology, artificial intelligence, and quantum mechanics continues to redefine the boundaries of modern science. As the proprietor of Quantum Mastery Industries and T3TInnovations (QMI, T3TInnovations), Eder leads a cutting-edge research and development company that pushes the limits of machine-to-brain integration, advanced AI systems, quantum computing, and the fusion of biological systems with computational models. His interdisciplinary expertise places him at the forefront of emerging scientific fields that will shape the future of humanity.

Holding both a PhD and an honorary DSc, Eder's academic achievements reflect his commitment to the pursuit of knowledge across multiple disciplines. His extensive involvement with pioneering institutions like Neuralink, OpenAI, and DARPA has propelled him to the centre of technological revolutions, particularly in areas of neuromodulation and the integration of biological and computational technologies. His innovations have laid the groundwork for advancements that bridge the gap between biology and artificial intelligence, setting new benchmarks for the future of human-machine symbiosis.

In addition to his advanced scientific work, Eder's technical proficiency shines through his operation of All Things Computational (ATC), a computer repair business that highlights his practical expertise. While

THE INFANCY OV THOUGHT

his research delves into the most complex realms of science, Eder remains grounded in real-world problem-solving, seamlessly merging theoretical understanding with technical application. His capacity for solving multifaceted problems, whether in the field of quantum mechanics, neurotechnology, or everyday computational systems, reflects the depth of his cognitive flexibility.

Eder's intellectual profile places him in the 99.999th percentile globally, with an IQ of 204 (± 16 z = 6.5 sd), far surpassing historical figures like Albert Einstein and John von Neumann. His near-perfect MCAT score of 524/528 further underscores his mastery across a range of scientific domains. Beyond raw intellectual power, his extraordinary creative abilities—particularly in writing—place him in the 99.99th percentile for linguistic and creative expression, as showcased in his literary works, including *The World In Prose, Vol. I*, *The World In Prose, Vol. II*, *The Chaos of Creation*, *A Prelude to the Fall* and *The Infancy Ov Thought*, (which he began writing at the age of fifteen).

Eder's contributions to theoretical physics, notably his novel equations that correct both the Rate of Expansion and the Rate of Entropy for the universe in relation to Hubble's Constant and Special Relativity, place him at the forefront of innovative scientific discovery. His theoretical advancements in quantum mechanics and astrophysics, along with his expertise in neuromodulation, solidify his status as a visionary scientist pushing the boundaries of what is possible.

His unparalleled intellectual and emotional intelligence allows him to navigate high-stress environments with poise and precision. Eder's exceptional emotional clarity, particularly in crisis scenarios, highlights his cognitive control and his adaptability in complex, real-time decision-making. His reflexivity and behavioural adaptability make him uniquely capable of handling diverse challenges, from life-threatening situations to intricate scientific puzzles.

Eder's visionary approach to science, his groundbreaking theoretical contributions, and his creative mastery position him as one of the leading intellectual figures of our time, blending rigorous scientific inquiry with philosophical depth and creative innovation.

$$z = \frac{204 - 100}{16} = 6.5$$

THE INFANCY OV THOUGHT

The Disseminative Divagation of the Godhead

The paradox of creation lies deep within the conceptualisation of a god who moulds man in his own likeness—a being imbued with consciousness, free will, volition, and a capacity for both innocence and cruelty. Yet, how can this paradox be reconciled with the existence of evil, with the depths of man's capacity for harm, and with the plight of the sentient beings who inhabit this mortal coil?

Why would a god, one who possesses infinite cognisance and mastery over the cosmos, grant entry to the unworthy—those who have wronged the innocent and the good? The enigma intensifies when we consider that a god with omnipresence, omnipotence, and omniscience should, by default, be a benevolent and beneficent force, no? Only by defining this god through the ethical logic of man's philosophy do we attempt to delineate the nature of the divine, be it linear, lineal, or nonlinear. Such definitions are humanity's feeble attempt to make sense of the incomprehensible. But in doing so, are we not simply attributing human qualities to a god that is beyond such limitations?

A god must, by necessity of thought, possess definable characteristics for the sake of discernment. If we are to understand, even in part, the nature of divinity, we must acknowledge certain qualities ascribed to him. And yet, even these traits—ageless, immortal, bornless, infinite, boundless—are inadequate in fully encapsulating the mystery of the Godhead. Consider, then, the Christian god of the Old Testament, an entity who embodies both light and darkness, love and hate, justice, and cruelty.

He must be ageless.

He must be immortal.

He must be bornless.

He must be aware.

THE INFANCY OV THOUGHT

He must be infinite.

He must be boundless.

He must intercede, and yet he must concede.

He must love, and he must hate.

He must be just, and he must be equal.

He must be formless.

And so, with this understanding of divine dualism—an encompassment of diffusive light and shadow—we move forward. Why, then, would God create man in his own image? Why nurture a being from infancy only to abandon him in adulthood? This is wilful neglect, is it not? A creator who brings forth life, then watches as it falters and fails, offers no salvation in the face of suffering. How can such negligence be reconciled with the notion of an all-powerful, loving god?

What of the concept of mortality? Can an ageless, immortal being truly understand the fleeting nature of human life, the pain of death, the suffering of the flesh? If Satan was right, and we have become as gods, possessing knowledge of good and evil, then what role does free will play? Have we not been burdened with the weight of our own awareness, our own self-consciousness?

And how strange, how ironic, that a book—penned by men but claimed to be inspired by the hand of the divine—would reference only the salvation of man. Is God idle, or is he idyllic? If hierarchies exist within the second heaven, how could God not expect many to be swayed by the romance of the dark, by the allure of Satan? Predisposition, predestination—how can free will exist within these bounds?

If God is truly boundless, then why the contradiction, the double standard? We are commanded to forgive our enemies, and yet we are told that Hell exists for those who have wronged God. Is he not held to the same standard? Should he not forgive those who have wronged him, just as we are expected to forgive those who trespass against us?

THE INFANCY OV THOUGHT

We are told that we are created in God's image, yet we are marred by a sinful nature. Is he not responsible for creating a faulty species? If he is omniscient and omnipotent, can he not wash away all iniquities, all sorrows? Why, then, does suffering persist? If God is not boundless, why is he credited with transcendent godhood? If we are subjected to the same fallen nature as animals, then why are the consequences of the fall so disparate between species? Why do American shorthair cats require meat to survive, while humans can live without it? Why do we not share the same conditions of the fall?

The inconsistencies are glaring. The constructs of God's love cannot hold when measured against the cruel realities of life. If God genuinely loves man more than the animals he created first, then why does he allow the slaughter of creatures like the elephant, an animal with a hippocampus far larger than that of a human, and thus capable of deeper memory and suffering? What justification is there for allowing man to rule over such creatures, to torture and hunt them for sport? The very idea is abhorrent.

What drives an artist? Spontaneity, perhaps. A passion ignited by witnessing the world, a spark that gives birth to creation. But what of the god who creates worlds? Does he not experience the same spontaneity, the same inexplicable drive? Or is his creation merely a reflection of his own nature—chaotic, boundless, unknowable?

The universality of intelligence dictates that all we experience is empirically real. And yet, there is an inherent falseness in the memory of it, a denial of the experiential. We are left grasping at shadows, trying to comprehend the incomprehensible, to reconcile the paradox of a god who is at once everything and nothing, boundless yet limited by his own creation.

And so, we are left with questions, endless questions, circling the divine like moths around a flame. Why create a species in your image, only to let it falter? Why allow suffering, cruelty, and injustice to persist, when you hold the power to erase it all? What is the purpose of such creation, if not to enlighten or uplift?

THE INFANCY OV THOUGHT

www.ingramcontent.com/pod-product-compliance
Lightning Source LLC
Chambersburg PA
CBHW020630220526
45464CB00001B/84